Experimental and Numerical Studies in Biomedical Engineering

Experimental and Numerical Studies in Biomedical Engineering

Special Issue Editors

Spiros V. Paras
Athanasios G. Kanaris

MDPI • Basel • Beijing • Wuhan • Barcelona • Belgrade

MDPI

Special Issue Editors
Spiros V. Paras
Aristotle University of Thessaloniki
Greece

Athanasios G. Kanaris
STFC
UK

Editorial Office
MDPI
St. Alban-Anlage 66
4052 Basel, Switzerland

This is a reprint of articles from the Special Issue published online in the open access journal *Fluids* (ISSN 2311-5521) from 2018 to 2019 (available at: https://www.mdpi.com/journal/fluids/special_issues/experimental_numerical_studies_biomedical_engineering)

For citation purposes, cite each article independently as indicated on the article page online and as indicated below:

LastName, A.A.; LastName, B.B.; LastName, C.C. Article Title. *Journal Name* **Year**, *Article Number*, Page Range.

ISBN 978-3-03921-247-7 (Pbk)
ISBN 978-3-03921-248-4 (PDF)

Contents

About the Special Issue Editors . . . . **vii**

Spiros V. Paras and Athanasios G. Kanaris
Experimental and Numerical Studies in Biomedical Engineering
Reprinted from: *Fluids* **2019**, *4*, 106, doi:10.3390/fluids4020106 . . . . **1**

Aleck H. Alexopoulos and Costas Kiparissides
A Computational Model for the Analysis of Spreading of Viscoelastic Droplets over Flat Surfaces
Reprinted from: *Fluids* **2018**, *3*, 78, doi:10.3390/fluids3040078 . . . . **4**

Angeliki T. Koupa, Yorgos G. Stergiou and Aikaterini A. Mouza
Free-Flowing Shear-Thinning Liquid Film in Inclined μ-Channels
Reprinted from: *Fluids* **2019**, *4*, 8, doi:10.3390/fluids4010008 . . . . **15**

Aikaterini A. Mouza, Olga D. Skordia, Ioannis D. Tzouganatos and Spiros V. Paras
A Simplified Model for Predicting Friction Factors of Laminar Blood Flow in Small-Caliber Vessels
Reprinted from: *Fluids* **2018**, *3*, 75, doi:10.3390/fluids3040075 . . . . **30**

Yorgos G. Stergiou, Athanasios G. Kanaris, Aikaterini A. Mouza and Spiros V. Paras
Fluid-Structure Interaction in Abdominal Aortic Aneurysms: Effect of Haematocrit
Reprinted from: *Fluids* **2019**, *4*, 11, doi:10.3390/fluids4010011 . . . . **43**

Stella K. Tsermentseli, Konstantinos N. Kontogiannopoulos, Vassilios P. Papageorgiou and Andreana N. Assimopoulou
Comparative Study of PEGylated and Conventional Liposomes as Carriers for Shikonin
Reprinted from: *Fluids* **2018**, *3*, 36, doi:10.3390/fluids3020036 . . . . **60**

Vigneswaran Narayanamurthy, Tze Pin Lee, Al'aina Yuhainis Firus Khan, Fahmi Samsuri, Khairudin Mohamed, Hairul Aini Hamzah and Madia Baizura Baharom
Pipette Petri Dish Single-Cell Trapping (PP-SCT) in Microfluidic Platforms: A Passive Hydrodynamic Technique
Reprinted from: *Fluids* **2018**, *3*, 51, doi:10.3390/fluids3030051 . . . . **76**

Dimosthenis Sarigiannis and Spyros Karakitsios
Advancing Chemical Risk Assessment through Human Physiology-Based Biochemical Process Modeling
Reprinted from: *Fluids* **2019**, *4*, 4, doi:10.3390/fluids4010004 . . . . **92**

Prodromos Arsenidis and Kostas Karatasos
Computational Study of the Interaction of a PEGylated Hyperbranched Polymer/Doxorubicin Complex with a Bilipid Membrane
Reprinted from: *Fluids* **2019**, *4*, 17, doi:10.3390/fluids4010017 . . . . **106**

About the Special Issue Editors

Spiros V. Paras received his Diploma in Chemical Engineering from Aristotle University of Thessaloniki (AUTh), Greece, he holds an MSc in Chemical Engineering from the University of Washington, Seattle, Wa, USA, and a Ph.D. in Chemical Engineering from AUTh, Greece. He is the Director of the Laboratory of Chemical Process and Plant Design, and Leader of the Process Equipment Design and Biomedical Engineering Group in the Department of Chemical Engineering in AUTh. He is currently Director of the Graduate Program "Chemical and Biomolecular Engineering". His research interests cover the subjects of multiphase flow in microfluidics, flows in biomedical applications, as well as high performance computing applications in science and fluid–structure interaction simulations. He has published numerous papers in peer-reviewed journals and has been a scientific reviewer for numerous international journals. Prof. Paras' teaching activities cover core chemical engineering subjects with an emphasis on:

- Multiphase Flows in Biomedical Applications and in Process Equipment;
- Chemical Plant Design;
- Medical Engineering;
- Advanced Measuring Techniques and CFD in Chemical Engineering.

Athanasios G. Kanaris received his Ph.D. in Chemical Engineering from Aristotle University of Thessaloniki, Greece, in 2008, focusing on the numerical and experimental studies of heat exchanger performance and design. He held a researcher position in the Department of Industrial Energy in Ecole des Mines de Douai, France, and worked as a Fluid Path Design engineer lead in Xaar, Cambridge, UK, where he oversaw and actively worked on the design and optimization of the ink distribution path inside a new state-of-the-art inkjet printhead during all crucial stages of development. His research interests cover the subjects of multiphase flow in microfluidics, heat transfer optimization design, flows in biomedical applications, as well as high performance computing applications in science, parallel computing optimization methods, fluid–structure interaction simulations and GPU acceleration in CFD/FEA applications. He is a Chartered Member of IMechE. He has been a scientific reviewer for more than 10 international journals and has participated in several research projects. Since 2017, he has been working as an HPC senior system administrator in the scientific computing department in the STFC Rutherford Appleton Laboratory (RAL) in the JASMIN project, providing and optimizing the scientific workflow for the research community.

fluids

MDPI

Editorial

Experimental and Numerical Studies in Biomedical Engineering

Spiros V. Paras [1,*] and Athanasios G. Kanaris [2]

[1] Department of Chemical Engineering, Aristotle University of Thessaloniki, 54124 Thessaloniki, Greece
[2] Scientific Computing Department, Rutherford Appleton Laboratory, Didcot OX11 0QX, UK; agkanaris@gmail.com
* Correspondence: paras@auth.gr; Tel.: +30-2310-996-174

Received: 3 June 2019; Accepted: 5 June 2019; Published: 6 June 2019

Keywords: microfluidics; blood flow; viscoelastic; falling film microreactor; μ-PIV; abdominal aortic aneurysm; hematocrit; computational fluid dynamics simulations; fluid–structure interaction; arterial wall shear stress; drug delivery; droplet spreading; passive trapping; cell capture; lab-on-a-chip; physiology-based biokinetics; liposomes; shikonin; human bio-monitoring

---

The term "biomedical engineering" refers to the application of the principles and problem-solving techniques of engineering to biology and medicine. Biomedical engineering is an interdisciplinary branch, as many of the problems health professionals are confronted with have traditionally been of interest to engineers because they involve processes that are fundamental to engineering practice. Biomedical engineers employ common engineering methods to comprehend, modify, or control biological systems, and to design and manufacture devices that can assist in the diagnosis and therapy of human diseases.

The goal of this Special Issue of *Fluids* is to be a forum for scientists and engineers from academia and industry to present and discuss recent developments in the field of biomedical engineering. It contains papers that tackle, both numerically (computational fluid dynamics studies) and experimentally, biomedical engineering problems, with a diverse range of studies focusing on the fundamental understanding of fluid flows in biological systems, modelling studies on complex rheological phenomena and molecular dynamics, design and improvement of lab-on-a-chip devices, modelling of processes inside the human body, and drug delivery applications. Contributions have focused on problems associated with subjects that include hemodynamical flows, arterial wall shear stress, targeted drug delivery, fluid–structure interaction/computational fluid dynamics (FSI/CFD) and multiphysics simulations, molecular dynamics modelling, and physiology-based biokinetic models.

In a comprehensive computational modelling study focused on complex rheological phenomena, Alexopoulos and Kiparissides [1] are using a macroscopic model to investigate the spreading of a linear viscoelastic fluid with changing rheological properties over flat surfaces. The computational model is based on a macroscopic mathematical description of the gravitational, capillary, viscous, and elastic forces. The dynamics of droplet spreading are determined in sessile and pendant configurations for different droplet extrusion or formation times for a hyaluronic acid solution undergoing gelation. The computational model is employed to describe the spreading of hydrogel droplets for different extrusion times, droplet volumes, and surface/droplet configurations. The effect of extrusion time is shown to be significant in the rate and extent of spreading.

In a series of studies where microfluidics engineering principles are used to improve understanding of biomedical phenomena, Paras and Mouza with their group contribute three different papers under this common theme:

Koupa et al. [2] present a study of the geometrical characteristics of a free-flowing non-Newtonian shear-thinning fluid flowing in an inclined open microchannel. The liquid film characteristics were measured by a non-intrusive technique that is based on the features of a micro particle image velocimetry (μ-PIV) system. Relevant computational fluid dynamics (CFD) simulations revealed that the volume average dynamic viscosity over the flow domain is practically the same as the corresponding asymptotic viscosity value, which can thus be used in the proposed design equations. A generalized algorithm for the design of falling film microreactors (FFMRs), containing non-Newtonian shear thinning liquids is also proposed.

Mouza et al. [3] present a simplified model for predicting friction factors of laminar blood flow in small-caliber vessels. The aim is to provide scientists with a correlation that can assist with the prediction of pressure drop that arises during blood flow in small-caliber vessels. This study has been conducted, like the previous one, using a combination of CFD simulations validated with relevant experimental data, acquired by the group. Experiments relate the pressure drop measurement during the flow of a blood analogue that follows the Casson model, that is, an aqueous glycerol solution that contains a small amount of xanthan gum. Results from this study lead to the proposal of a simplified model that incorporates the effect of the blood flow rate, the hematocrit value (35–55%) and the vessel diameter (300–1800 μm) and predicts, with satisfactory accuracy, pressure drop during laminar blood flow in healthy small-caliber vessels.

Stergiou et al. [4] in their contribution incorporated a complex multiphysics simulation to provide a realistic model of blood flow and to numerically examine, using a fully coupled fluid–structure interaction (FSI) method, the complicated interaction between the blood flow and the abdominal aortic aneurysm (AAA) wall. The study investigates the possible link between the dynamic behavior of an AAA and the blood viscosity variations attributed to the haematocrit value, while it also incorporates the pulsatile blood flow, the non-Newtonian behavior of blood and the hyperelasticity of the arterial wall. Results in terms of wall shear stress (WSS) show that its fluctuations due to haematocrit changes can alter the mechanical properties of the arterial wall and increase the growth rate of the aneurysm or even its rupture possibility.

In the field of drug delivery, Tsermentseli et al. [5] present a comparative study between PEGylated and conventional liposomes, as carriers for shikonin. Liposomes are considered one of the most successful drug delivery system. On the other hand, shikonin and alkannin, a pair of chiral natural naphthoquinone compounds, are widely used due to their various pharmacological activities. The study reports the effects of different lipids and polyethylene glycol (PEG) on parameters related to particle size distribution, polydispersity index, ζ-potential, drug-loading efficiency and stability of the prepared liposomal formulations. Three types of lipids were assessed (DOPC, DSPC, DSPG), separately and in mixtures, forming anionic liposomes with good physicochemical characteristics, high entrapment efficiencies, satisfactory in vitro release profiles, and good physical stability. The shikonin-loaded PEGylated sample with DOPC/DSPG, demonstrated the most satisfactory characteristics and is considered promising for further design and improvement of these type of formulations.

In the field of lab-on-a-chip research, Narayanamurthy et al. [6] present a study on pipette Petri dish single-cell trapping (PP-SCT) as an application of a passive hydrodynamic technique. PP-SCT is simple and cost-effective with ease of implementation for single cell analysis applications. In their study, passive microfluidic-based biochips capable of vertical cell trapping with the hexagonally-positioned array of microwells are exhibited and a wide operation at different fluid flow rates of this novel technique is demonstrated. Using human lung cancer cells, single-cell capture (SCC) capabilities of the microfluidic-biochips are found to be improving from the straight channel, branched channel, and serpent channel, accordingly. Multiple cell capture (MCC) is on the order of decreasing from the straight channel, branch channel, and serpent channel. Among the three designs investigated, the serpent channel biochip offers high SCC percentage with reduced MCC and NC (no capture) percentage. Using the PP-SCT technique, flow rate variations can be precisely achieved.

In a study focusing on the use of physiology-based biokinetic (PBBK) models, Sarigiannis and Karakitsios [7] aim at the development of a lifetime PBBK model that covers a large chemical space, which, when coupled with a framework for human biomonitoring (HBM) data assimilation, provides an advanced chemical risk assessment method. The methodology developed was demonstrated in the case of bisphenol A (BPA), where exposure analysis was based on European HBM data. Based on their calculations, it was found that current exposure levels in Europe are below the temporary tolerable daily intake (t-TDI) proposed by the European Food Safety Authority (EFSA). The authors propose refined exposure metrics, which show that environmentally relevant exposure levels are below the concentrations associated with the activation of biological pathways relevant to toxicity.

Finally, in a computational study using molecular dynamics (MD), Arsenidis and Karatasos [8] present fully atomistic MD simulations employed to study the interactions between a complex comprised by a PEGylated hyperbranched polyester (HBP) and doxorubicin molecules, with a model membrane in an aqueous environment. The effects of the presence of the lipid membrane in the drug molecules' spatial arrangement are examined in detail and the nature of their interaction with the latter are discussed and quantified where possible. A partial migration of the drug molecules towards the membrane's surface takes place, while clustering behavior of the drug molecules appeared to be enhanced in the presence of the membrane, and development of a charge excess close to the surface of the hyperbranched polymer and of the lipid membrane is observed. The build-up of the observed charge excesses, together with the changes in the diffusional behavior of the drug molecules are of particular interest, regarding the latest stages of the liposomal disruption and the release of the cargo at the targeted sites.

We would like to thank the contributors to this Special Issue for sharing their research, and the reviewers for generously donating their time to select and improve the manuscripts.

Conflicts of Interest: The authors declare no conflict of interest.

References

1. Alexopoulos, A.H.; Kiparissides, C.A. Computational Model for the Analysis of Spreading of Viscoelastic Droplets over Flat Surfaces. *Fluids* **2018**, *3*, 78. [CrossRef]
2. Koupa, A.T.; Stergiou, Y.G.; Mouza, A.A. Free-Flowing Shear-Thinning Liquid Film in Inclined μ-Channels. *Fluids* **2019**, *4*, 8. [CrossRef]
3. Mouza, A.A.; Skordia, O.D.; Tzouganatos, I.D.; Paras, S.V. A Simplified Model for Predicting Friction Factors of Laminar Blood Flow in Small-Caliber Vessels. *Fluids* **2018**, *3*, 75. [CrossRef]
4. Stergiou, Y.G.; Kanaris, A.G.; Mouza, A.A.; Paras, S.V. Fluid-Structure Interaction in Abdominal Aortic Aneurysms: Effect of Haematocrit. *Fluids* **2019**, *4*, 11. [CrossRef]
5. Tsermentseli, S.K.; Kontogiannopoulos, K.N.; Papageorgiou, V.P.; Assimopoulou, A.N. Comparative Study of PEGylated and Conventional Liposomes as Carriers for Shikonin. *Fluids* **2018**, *3*, 36. [CrossRef]
6. Narayanamurthy, V.; Lee, T.P.; Khan, A.Y.F.; Samsuri, F.; Mohamed, K.; Hamzah, H.A.; Baharom, M.B. Pipette Petri Dish Single-Cell Trapping (PP-SCT) in Microfluidic Platforms: A Passive Hydrodynamic Technique. *Fluids* **2018**, *3*, 51. [CrossRef]
7. Sarigiannis, D.; Karakitsios, S. Advancing Chemical Risk Assessment through Human Physiology-Based Biochemical Process Modeling. *Fluids* **2019**, *4*, 4. [CrossRef]
8. Arsenidis, P.; Karatasos, K. Computational Study of the Interaction of a PEGylated Hyperbranched Polymer/Doxorubicin Complex with a Bilipid Membrane. *Fluids* **2019**, *4*, 17. [CrossRef]

fluids

MDPI

Article

A Computational Model for the Analysis of Spreading of Viscoelastic Droplets over Flat Surfaces

Aleck H. Alexopoulos [1] **and Costas Kiparissides** [1,2,*]

1 Chemical Process & Energy Resources Institute, 6th km Harilaou-Thermi rd., P.O. Box 60361, Thessaloniki 57001, Greece; aleck@cperi.certh.gr

2 Department of Chemical Engineering, Aristotle University of Thessaloniki, Thessaloniki 54124, Greece

* Correspondence: costas.lpre@cperi.certh.gr; Tel.: +30-2310-498161

Received: 7 September 2018; Accepted: 19 October 2018; Published: 22 October 2018

Abstract: The spreading of viscous and viscoelastic fluids on flat and curved surfaces is an important problem in many industrial and biomedical processes. In this work the spreading of a linear viscoelastic fluid with changing rheological properties over flat surfaces is investigated via a macroscopic model. The computational model is based on a macroscopic mathematical description of the gravitational, capillary, viscous, and elastic forces. The dynamics of droplet spreading are determined in sessile and pendant configurations for different droplet extrusion or formation times for a hyaluronic acid solution undergoing gelation. The computational model is employed to describe the spreading of hydrogel droplets for different extrusion times, droplet volumes, and surface/droplet configurations. The effect of extrusion time is shown to be significant in the rate and extent of spreading.

Keywords: spreading; gelation; hydrogel; hyaluronic; viscoelastic; viscous; gravitational; capillary

1. Introduction

The spreading of viscous and viscoelastic droplets over surfaces is an immensely important subject that has been studied extensively [1–4]. The detailed computational description is inherently an extremely challenging problem as it involves complex flow fields, movement and deformation of the droplet interface as well as creation of a new interface and these processes alter the boundary to the computational domain. Fluid viscoelasticity further complicates the problem and can manifest in many ways, e.g., memory, elastic forces, yield-stress effects.

To date computational techniques to solve these types of problems have been developed following a case-by-case problem-oriented approach, e.g., Reference [5]. On the other hand, commercial Computation Fluid Dynamics, CFD, products until recently have lacked efficient boundary updating for many of the large deformation problems. Currently, commercial products based on Finite Element Method, FEM, techniques with well-structured grids appear to be the most accurate in terms of surface updating during large deformations [6]. Continuing efforts to develop computational techniques for such challenging programs include many hybrid approaches, e.g., boundary element methods for the interface and FEM for the internal nonlinear viscoelastic or inertial terms [7]. These and other approaches require significant expertise and further testing before they can be adopted by the general scientific community.

At the same time there is significant amount of experimental investigations involving free-surface flows of viscous or viscoelastic fluids. Spreading of droplets over flat surfaces is a very common laboratory test to determine "spreadability" of fluids, (e.g., with lubricant oils [8] and food products [9]) as well as an industrial test (e.g., with foamed cement [10]). The droplet spreading tests can become even more difficult to interpret when considering viscoelastic fluids, curved surfaces, or time-variable fluid properties. Because of these difficulties there is a need to provide a simple means to describe spreading tests of complex fluids in order to extract more meaningful data as well as to obtain a better understanding of more complex coating processes.

In this work we develop a simple droplet spreading model for a linear viscoelastic fluid and connect it to nonlinear processes that affect the rheological properties of the fluid. The literature is currently lacking in such simple models. These models can be employed to facilitate the interpretation of simple spreading tests and potentially as a design tool for many processes involving spreading and time varying fluid properties. In the section that follows the computational model is described building on the work of Härth and Schubert [11]. Next the specific hyaluronic acid system which forms droplets examined in this work is described. In the third section the model is tested with the spreading of viscoelastic droplets with time-varying properties.

2. Droplet Spreading Model

To develop a simple macroscopic film spreading model the macroscopic model of Härth and Schubert [11] is employed which considers viscous, gravitational, and surface forces for partially or fully wetting droplets. The model is extended to include elastic forces, consider pendant drop configurations in addition to sessile drop, and is applied to a gelling system with time-varying rheological properties. It should be noted that the spherical cap approximation is realistic if the initial droplet radius is less than the capillary length, $L_C = \sqrt{\gamma_L/\rho\, g}$, or the Bond number, Bo, is less than one [12]:

$$Bo = \frac{\Delta\rho\, g\, R_0^2}{\gamma_L} = \left(\frac{R_0}{L_C}\right)^2 < 1 \tag{1}$$

where ρ is the fluid density, g, is the gravitational acceleration constant, γ_L is the fluid surface energy, and R_0 is the initial curvature at the apex which is equal to the initial spherical cap radius.

The basic geometry of a spherical-cap droplet spreading on a flat surface is shown in Figure 1.

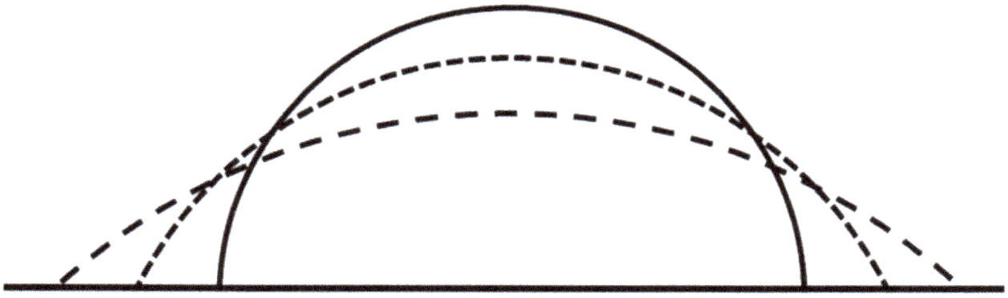

Figure 1. Drop shapes during spreading. Initial spherical cap (**solid line**), transition shape (**short-dashed line**), steady state "pancake" shape (**long-dotted line**).

For a spherical cap droplet of base radius, r, and height, h, the volume, V, is given by:

$$V = \frac{\pi}{6}\left(3\, r^2\, h + h^3\right) \tag{2}$$

and it is constant with time as long as there are no physical or chemical changes in the fluid and as long as there are no mass losses, e.g., due to evaporation.

Consequently the differential in height is given by [13]:

$$dh = -\frac{2\, r\, h}{r^2 + h^2} dr \tag{3}$$

2.1. Forces Acting on Droplet during Spreading

The total force, F, acting in the radial direction is the sum of capillary, viscous, gravitational, and elastic terms. To determine these forces a macroscopic approach is followed assuming flat droplets, i.e., $h << r$, and small Bond numbers, i.e., $Bo < 1$. The forces are determined by considering the various contributions to the droplet energy, E, during an infinitesimal spreading step of dr and dh (Figure 2) during which the total radial force, F, is given as:

$$F = -\frac{dE}{dr} = -\frac{\partial E}{\partial r} + \frac{\partial E}{\partial h}\frac{2rh}{r^2 + h^2} \tag{4}$$

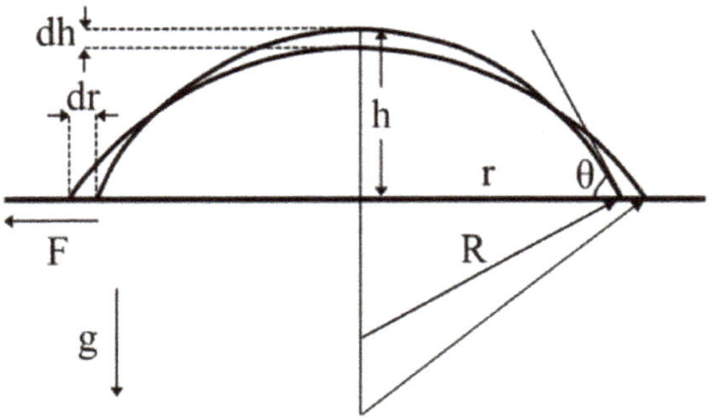

Figure 2. Spreading of droplet over a flat surface over a time period of dt.

With these assumptions it was shown in [11] that the capillary force, F_C, is given by:

$$F_C = 2\,\pi\,r\left(S + \gamma_L\,\frac{2\,r^2}{r^2 + h^2}\right) \tag{5}$$

where S is the spreading coefficient given by:

$$S = \gamma_S - \gamma_{SL} - \gamma_L \tag{6}$$

where γ_S is the surface energy of the solid and γ_{SL} is the surface/fluid interface energy.

Following Härth and Schubert, and by considering the potential energy of a spherical cap as an integral over horizontal slabs of thickness dz:

$$E = \int_0^h \rho g z dV \tag{7}$$

the gravitational force, F_G, is determined to be [11]:

$$F_G = \rho\,g\,\pi\,h^2\,\frac{r}{3}\left(\frac{r^2}{r^2 + h^2}\right) = \rho\,g\,\frac{\pi}{6}\,\frac{r^3\,h}{R} \tag{8}$$

where R is the radius of the spherical cap (Figure 2) which is equal to:

$$R = \frac{r^2 + h^2}{2\,h} \tag{9}$$

The viscous force, F_V, can only be approximated in a macroscopic approach because of the unknown velocity profile, e.g., adjacent to the contact line. The movement of the contact line can be very complicated and dynamic. The contact line does not always move smoothly (e.g., stick-slip motion, see [1]) and is not always well defined (e.g., the fluid over the contact line can move over a thin layer of air). When considering microscopic effects van der Waals forces [14] and line tension effects [15] can become important and nanoscale effects require different considerations [16].

The viscous force for a Newtonian fluid undergoing simple shear flow is proportional to a shear stress, τ, multiplied by a surface area, A, parallel to the direction of flow according to:

$$F_V = \tau\, A \;=\; \eta\, \dot{\gamma}\, A \tag{10}$$

where η is the viscosity and $\dot{\gamma}$ is the shear rate. Here it is assumed that the dominating nature of flow is simple shear as a stick boundary condition can be assumed for most of the contact area of the droplet. This assumption over-estimates the shear rate only in a small region near the moving contact line which during the slip transition does not flow via simple shear. Consequently, we have:

$$F_V = \eta\, \frac{\dot{r}}{w}\, 2\pi r w = 2\, \pi\, r\, \eta\, \dot{r} \tag{11}$$

where w is the average height of the droplet. It should be noted that in Härth and Schubert the viscous force (without the 2π term) was adapted to:

$$F_V = r\, \eta\, \dot{r} \;\approx\; \frac{r^6\, \eta\, \dot{r}}{\xi\, V^2} \tag{12}$$

where ξ = 37.1 m^{-1} is considered a universal constant.

In this work the elastic contribution of a linear viscoelastic fluid of the Maxwell type:

$$\tau + \lambda\, \frac{d\tau}{dt} = \eta\, \dot{\gamma} \tag{13}$$

where η is the Maxwell viscosity and the relaxation time λ is given by

$$\lambda \;=\; \frac{\eta}{E} \tag{14}$$

where E is the elasticity of the Maxwell fluid.

In order to determine the elastic contribution from linear viscoelastic Equation (14) it is clear that there should be a first-order relaxation term $e^{-\frac{t}{\lambda}}$ and that the average elastic stress should be proportional to the average relative deformation, $\varepsilon(t)$, and the elasticity E, according to:

$$\tau_E(t) \;=\; E\, \varepsilon(t)\, e^{-\frac{t}{\lambda}} \tag{15}$$

The average relative deformation is approximated by the deviation from the initial spherical shape so that:

$$\varepsilon(t) \approx \frac{r(t) - R_0}{R_0} \tag{16}$$

Following the same procedure as with the viscous force we obtain the elastic force term for a Maxwell fluid:

$$F_{E,M} = 2\, \pi\, r\, h\, \frac{\eta}{\lambda}\, \frac{r - R_0}{R_0}\, e^{-t/\lambda} \tag{17}$$

The net driving force, F_{tot}, in the radial direction for deformation and spreading of a viscoelastic droplet on a flat surface is a sum of capillary, gravitational, viscous, and elastic terms.

$$F_{tot} = F_C + F_G - F_V - F_{E,M} \tag{18}$$

2.2. Spreading Model

For highly viscous or viscoelastic fluids a quasi-steady state assumption is valid in which the net acceleration is much smaller than the other processes. The rate of change in the radius, dr/dt, can then be obtained from:

$$0 \approx 2\,\pi\,r\left(S+\gamma_L\,\frac{2\,r^2}{r^2+h^2}\right)+\rho\,g\,\frac{\pi}{6}\,\frac{r^3\,h}{R}-2\,\pi\,r\,\eta\,\frac{dr}{dt}-2\,\pi\,r\,h\,\frac{\eta}{\lambda}\,\frac{r-R_0}{R_0}\,e^{-t/\lambda} \tag{19}$$

Consequently, Equation (19) can be solved for $\frac{dr}{dt}$ to obtain:

$$\frac{dr}{dt} = \frac{S}{\eta}+\frac{\gamma_L}{\eta}\,\frac{r^2}{h\,R}+\frac{1}{12}\,\frac{\rho\,g}{\eta}\,\frac{r^2\,h}{R}-\frac{h}{\lambda}\,\frac{r-R_0}{R_0}\,e^{-t/\lambda} \tag{20}$$

The above equation is solved together with:

$$\frac{dh}{dt} = -\frac{2\,r\,h}{r^2+h^2}\,\frac{dr}{dt} \tag{21}$$

which is obtained from Equation (3) together with Equation (9) for R in order to provide the time variation of the radius of contact, r, height, h, and aspect ratio, $Z = r/h$. If the initial contact radius $r_0 = r(0)$ is known for a given droplet volume V then, from Equation (2), the following cubic equation is solved for the initial height of the spherical cap, $h_0 = h(0)$:

$$h(0)^3+3\,r(0)^2\,h(0)-\frac{6\,V}{\pi} = 0 \tag{22}$$

Setting $x = r/R_0$ and $y = h/R_0$ and dividing by R_0 we have:

$$\frac{dx}{dt} = \left[\frac{S}{\eta\,R_0}\right]+\left[\frac{\gamma_L}{\eta\,R_0}\right]\frac{2\,x^2}{x^2+y^2}+\frac{1}{6}\left[\frac{\rho\,gR_0}{\eta}\right]y^2\,\frac{x^2}{x^2+y^2}-\left[\frac{1}{\lambda}\right]y\,(x-1)\,e^{-t/\lambda} \tag{23}$$

where the terms in square brackets have units of 1/s.

Equation (23) is solved together with Equation (21) in the following form:

$$\frac{dy}{dt} = -\frac{2\,x\,y}{x^2+y^2}\,\frac{dx}{dt} \tag{24}$$

Selecting a characteristic time of $t^* = \sqrt{R_0/g}$ we can obtain the following dimensionless forms:

$$\frac{dx}{d\tau} = \frac{\sigma}{Ca}+\frac{1}{Ca}\,\frac{2\,x^2}{x^2+y^2}+\frac{1}{6}\,\frac{Bo}{Ca}\,y^2\,\frac{x^2}{x^2+y^2}-\frac{1}{De}\,y\,(x-1)\,e^{-\tau/De} \tag{25}$$

and

$$\frac{dy}{d\tau} = -\frac{2\,x\,y}{x^2+y^2}\,\frac{dx}{d\tau} \tag{26}$$

where $\tau = t/t^*$, $\sigma = S/\gamma_L$, Ca is the Capillary number, $Ca = \mu\,\sqrt{\rho g}/\gamma_L$, and De is the Deborah number, $De = \lambda/t^*$.

The effect of inverted droplets (i.e., pendant droplets) hanging from a flat surface can be studied by changing the sign in the gravitational term of Equations (23) and (24) or the dimensionless Equations (25) and (26).

2.3. Varying Rheological Properties

Rheological properties can change with time due to physical (e.g., compositional changes due to evaporation) and chemical (e.g., reaction) processes. These changes are reflected in rheological measurements, e.g., oscillatory rheometry, leading to time varying storage (i.e., G') and loss (i.e., G'') moduli of the fluid.

In order to describe the deformation of a viscoelastic hydrogel droplet undergoing gelation a simple linear viscoelastic model with time varying material properties was considered. Note that the Maxwell fluid element converges to Newtonian when $E \to \infty$ as the viscosity pot and the spring are in series. The loss and storage moduli data, at a specific frequency, ω, can be related to the Maxwell fluid coefficients according to [11]:

$$E = G'' \left[\left(\frac{G'}{G''} \right)^2 + 1 \right] / \left(\frac{G'}{G''} \right) \tag{27}$$

and

$$\eta = \lambda \ = \frac{G'}{G''} \frac{}{\omega} \tag{28}$$

It should be noted that more complicated rheological models require additional rheological data to be properly characterized and cannot be easily decomposed into viscous and elastic component as in this simple analysis.

2.4. System Studied

The system studied consists of an enzymatically crosslinking hyaluronic acid (HA) system [17]. Specifically, Lee et al. [17] provide results for oscillatory rheometry experiments which were performed while HA-tyramine hydrogel was formed via crosslinking of tyramine moieties catalyzed by hydrogen peroxide (H_2O_2) and horseradish peroxidase (HRP). The oscillatory rheometry results for the loss and storage moduli (obtained with a constant deformation of 1% at 1 Hz and at a temperature of 37 °C) are summarized in Table 1. Note that for this specific system (i.e., with 728 mM of H_2O_2 and 0.025 units per ml of HRP) the gel point [18] where the hydrogel transitions from a viscoelastic liquid to a viscoelastic solid occurs at 48 s.

Table 1. Loss and storage moduli (extracted from [11]).

t, s	*G'*, Pa	*G''*, Pa
0	0.8	4.2
25	3.6	11
48	28.4	28.4
80	150	42
100	285	41
150	720	31
200	1120	23
250	1400	19
300	1850	15
400	2350	15
600	2800	15

2.5. Physical Model and Simulation Algorithms

As a test system an extrusion syringe was considered where a droplet is directed to a flat surface either facing upwards (i.e., sessile configuration) or downwards (i.e., pendant configuration). The gelling fluid is extruded onto the surface where it forms an initial half droplet. The syringe is retracted to allow the droplet to spread freely. It is assumed that the HA solution is mixed instantaneously and completely at the beginning of the syringe and gelation continues throughout the extrusion and spreading processes. Figure 3 displays a typical setup in pendant configuration.

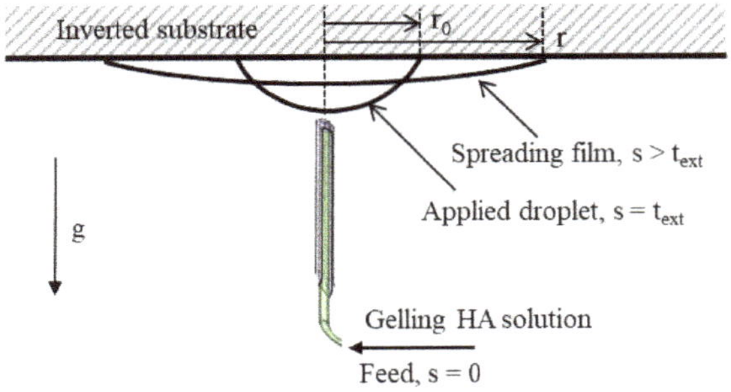

Figure 3. Film application and spreading onto an inverted substrate. s = time from inflow to syringe = $t + t_{ext}$. Flow rate Q ~1–10 cm^3/min.

Early experimental studies (not reported here) indicate that a critical property for spreading and film formation is the extrusion time, t_{ext}, or the time required to form the initial droplet especially for rapidly gelling systems. If the extrusion times are too large no spreading is observed of the droplet and there is no film formation. Large droplets were found to detach easily especially with low viscosity droplets, i.e., short extrusion times.

In this work a computational including the residence time of the gelling HA solution in the syringe is taken into account. The known geometric properties of the system are the droplet volume and the initial contact radius. The known physical properties are the density, surface tension, viscosity, spreading coefficient, and the rheological properties of the gelling HA solution, i.e., loss and storage moduli.

The simulation procedure is shown in Figure 4. The simulation begins with solution of the cubic Equation (22) for the droplet height. Next, the spreading equations, i.e., (25) and (26) are solved with time. At each time step the total time of gelation, s, and of spreading, t, are determined. Based on the rheology data of Lee et al. [17] the rheological parameters λ and η are calculated at the corresponding gelation time, s. Simulations proceed until the net change per time step becomes less than a limit value.

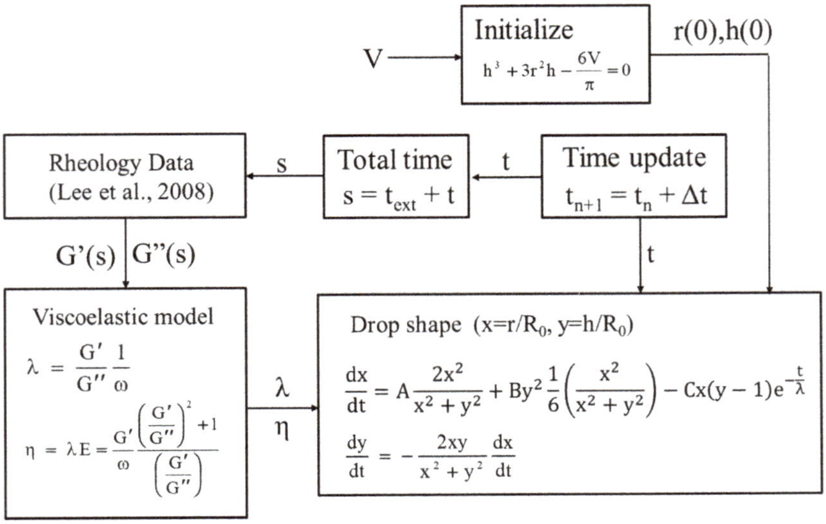

Figure 4. Film spreading algorithm for linear viscoelastic fluids with variable properties.

In this work the Maxwell model is used as a simple representation of a linear viscoelastic fluid. This approach can be implemented with more complex linear viscolelastic models. The drop shape algorithm assumes spherical cap shapes which has been shown to agree with experimental data for broad "pancake" shaped droplets. The total time "s" is employed in the rheology model and includes the extrusion time. In this way the extrusion model is connected to the spreading model. For example, if the extrusion rate is very slow (or the extrusion time is very large) then the extruded droplets will be too viscous and elastic to adequately spread and will detach instead. It should be noted that for Newtonian fluids the computational model reduces to a model similar to that of Härth and Schubert which was validated for the spreading of sessile Newtonian fluid droplets [11].

3. Results

For the gelling hyaluronic system studied in this work a density difference of $\Delta\rho = 103$ Kg/m^3 and surface tensions of $\gamma = 15$ and 45 mN/m were assumed. Also, the initial contact radii and the droplet volumes ranged between 0.2–1 cm and 2–4 cm^3, respectively. The fluid, i.e., gelling hyaluronic acid solution, was assumed to fully wet the surface (i.e., $S = 0$) and to form a droplet after an extrusion time of t_{ext}, Various extrusion times from 10 to 120 s were examined. Both flat upward-facing and inverted geometries corresponding to sessile ($g > 0$) and pendant ($g < 0$) configurations for the initial droplet were considered.

The results are shown in Table 2 in terms of the final contact radius, r, height, h, droplet radius, R_c, and the spreading aspect ratio $Z = r/h$. For sessile and pendant droplets, the spreading process results in an exponentially decreasing contact line velocity. As expected the surface tension plays an important role. The spreading aspect ratio for case 1 ($\gamma_L = 15$ mN/m) was $Z = 7.7$ and for case 3 (i.e., $\gamma_L = 4$ mN/m) it was nearly four times larger at $Z = 27.7$.

The effect of droplet formation or extrusion time was also examined. From the results in Table 2 it is clear that delaying the film spreading (by increasing the application or extrusion time) changes the rheological properties of the gel to such a point that the elastic forces inhibit spreading. As the extrusion time increased from $t_{ext} = 10$ to 60 s the spreading aspect ratio decreased from $Z = 27.7$ to 2.9. If the extrusion time is increased further, then the gel does not spread at all despite it wetting the surface.

Table 2. Simulation Results.

γ mN/m	g m/s^2	$r(0)$ cm	V cm^3	t_{ext} s	r cm	h cm	R_c cm	$Z = r/h$
15	9.8	1	4	30	2.71	0.35	9.5	7.7
45	9.8	1	4	10	4.16	0.15	59.0	27.7
45	9.8	1	4	20	3.61	0.20	33.6	18.1
45	9.8	1	4	30	3.16	0.25	19.7	12.6
45	9.8	1	4	40	2.66	0.36	10.0	7.4
45	9.8	1	4	50	2.27	0.49	5.5	4.6
45	9.8	1	4	60	1.92	0.66	3.1	2.9
45	9.8	1	4	90	1.31	1.17	1.3	1.1
45	9.8	1	4	120	1.09	1.39	1.1	0.8
45	9.8	1	1	30	2.15	0.13	16.8	16.6
45	−1.0	1	4	30	2.78	0.33	11.9	8.4
45	−2.0	1	4	30	2.58	0.38	9.0	6.8
45	−3.0	1	4	30	0	0	-	-
45	−3.0	1	4	10	0	0	-	-
45	−3.0	0.2	1	30	13.51	0.43	58.2	31.4
45	−9.8	1	4	30	0	0	-	-
45	−9.8	1	1	60	1.16	0.44	1.7	2.6
45	−9.8	1	1	30	2.01	0.16	13.0	12.6
45	−9.8	1	1	20	2.43	0.11	23.4	22.1

Because the Bond number is small, the effect of inverted or pendant droplets can be studied by changing the sign in the gravitational term in Equation (26) which in Table 2 is denoted with negative gravity. It is clear that stable pendant droplets can be obtained below a specific mass and these display spreading but to a smaller degree than the corresponding sessile droplets of the same mass.

From Equation (25) it can be shown that the spreading pendant films are stable, i.e., spreading and not retracting, when $h_0 = h(0) < 3.46\ L_C$ or, in terms of the Bond number defined using the initial height of the film h_0:

$$Bo = \frac{\Delta\rho\ g\ h_0^2}{\gamma_L} < 12 \tag{29}$$

Note that when pendant drops are retracting they either form a stable pendant droplet or detach. Both of these outcomes are beyond the scope of the current macroscopic model which examines only stable, i.e., $Bo < 1$, and flat, i.e., $h/r < 1$, droplets.

In Figure 5 the spreading of hydrogel droplets on an inverted surface is shown. It is clear that the final aspect ratio, $Z = r/h$, depends strongly the initial contact radius of the droplet as well as the extrusion time due to the gelling reaction and changing rheological properties. The dashed line separates the spreading and retracting regimes in terms of the initial contact radius according to Equations (2) and (29). Note that with the current model metastable configurations are possible in the retracting regime, i.e., for $r_0 < 1.237$ cm, where the droplet is unstable but kinetically frozen.

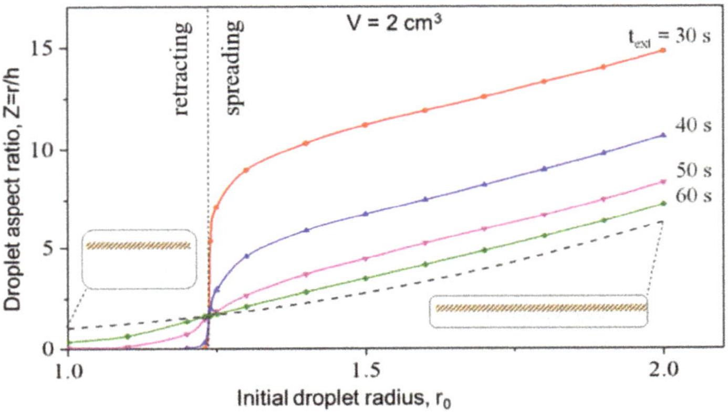

Figure 5. Initial (dashed line) and final (full line) droplet aspect ratios, i.e., $Z = r/h$, for different extrusion times and initial droplet contact radii, r_0. ($V = 2\ cm^3$, $\gamma = 45$ mN/m).

In Figure 6 it can be seen that for a given droplet volume (i.e., $V = 2\ cm^3$) increasing the extrusion time results in decreased droplet spreading due to increased elasticity and viscosity. This reflects the coupling of the applicator syringe and droplet spreading problem. Clearly very slow application speeds will result in inadequate spreading.

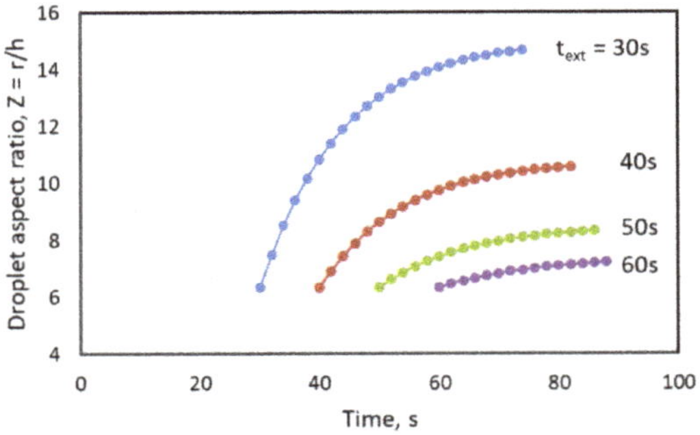

Figure 6. Droplet shape as a function of total time, s, for different extrusion times. ($V = 2\ cm^3$, $\gamma = 45$ mN/m).

In Figure 7 the effect of elasticity on droplet spreading is shown. The elasticity was increased by ×3 and ×8 times compared to the normal case with an extrusion time of 30 s. The effect of elasticity alone is to significantly decrease the response time and the extent of spreading. This is expected to be a general effect even with other—reasonable—rheological models.

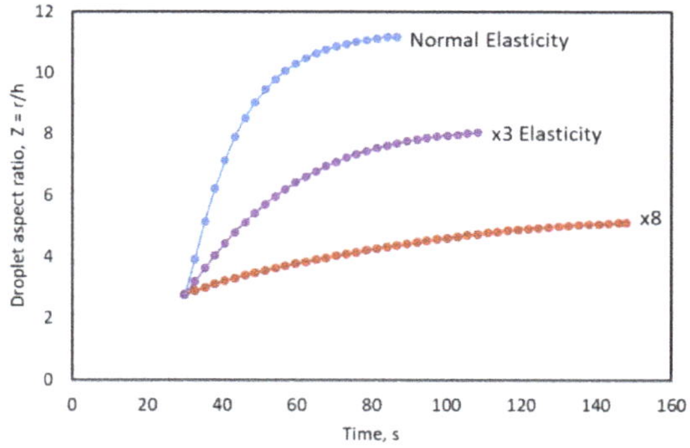

Figure 7. Effect of elasticity on droplet spreading. t_{ext} = 30 s (V = 2 cm^3, γ = 45 mN/m).

4. Discussion and Conclusions

In this work the model of Härth and Schubert [11] is extended to account for linear viscoelastic effects. The model is employed to describe both sessile and pendant drop configurations in a case study with spreading of gelling hyaluronic acid solution. The effect of gelling time and elasticity are shown to be significant in terms of the extent of spreading of the gelling hyaluronic acid solution.

The computational approach described in this work can be implemented with other simple linear viscoelastic models. The approach could be extended beyond spherical caps, e.g., to ellipsoidal caps, as a more general approximation of sessile and pendant droplets. Furthermore, limitations could be placed on the moving contact angles which are determined by $sin\theta = r/R$. More realistic descriptions of changes on the contact angle at the moving contact line necessitate some type of model of the underlying surface which can be a challenging task. For example, soft deformable surfaces can display significant deformation at the contact line [19].

Application of hydrogel or other viscoelastic fluids by spreading over biological membranes and mucous layers and other surfaces remains a very complex subject. A complicating factor is when the spreading fluid is undergoing chemical or physical changes resulting in variable rheological properties with time. This type of a dynamic system may be a desirable solution when spreading occurs while the fluid is still of relative low viscosity but, after spreading has finished, rheological changes such as increased viscosity and mucoadhesiveness help to keep the film in place.

This work demonstrates a simple model providing a connection between dynamic rheological data and droplet spreading behavior. Although many assumptions are made, the model can describe the spreading behavior over flat surfaces, including the effect of viscoelasticity. Approximate models can also help in designing and interpreting spreading experiments of complex viscoelastic fluids. Approximate models can help reduce the design space of a system involving several elements such as mixing, delivery, spreading. The potential exists to employ these types of simple models to more complex hydrogel delivery systems, e.g., via droplet sprays, other gelling mechanism, e.g., thermos-reversible, and other geometries, e.g., slightly curved surfaces. Of the many possible improvements to the model, including the interaction of the film with the substrate, beyond the simple surface energy of Equation (4), seems the most interesting and can be achieved by considering mucoadhesion models or including mechanisms for contact line motion dynamics.

Author Contributions: Conceptualization, A.A. and C.K.; Methodology, A.A.; Software, A.A.; Validation, A.A.; Formal Analysis, A.A.; Investigation, A.A. and C.K.; Resources, C.K.; Data Curation, A.A.; Writing-Original

Draft Preparation, A.A.; Writing-Review & Editing, A.A.; Visualization, A.A.; Supervision, A.A. and C.K.; Project Administration, C.K.; Funding Acquisition, C.K.

Funding: The present research has been financially supported by EU under the European Framework Programme for Research and Innovation Horizon 2020 (Grant No. 721098).

Conflicts of Interest: The authors declare no conflict of interest. The funders had no role in the design of the study; in the collection, analyses, or interpretation of data; in the writing of the manuscript, and in the decision to publish the results.

References

1. Dussan, E.B. On the spreading of liquids on solid surfaces: Static and dynamic contact lines. *Ann. Rev. Fluid Mech.* **1979**, *11*, 371–400. [CrossRef]
2. Uppal, A.S.; Craster, R.V.; Matar, O.K. Dynamics of spreading thixotropic droplets. *J. Non-Newton. Fluid Mech.* **2017**, *240*, 1–14. [CrossRef]
3. Xu, H.; Clarke, A.; Rothstein, J.P.; Poole, R.J. Viscoelastic drops moving on hydrophilic and superhydrophobic surfaces. *J. Colloid Interface Sci.* **2018**, *513*, 53–61. [CrossRef] [PubMed]
4. Bird, R.B.; Armstrong, R.C.; Hassager, O. *Dynamics of Polymeric Liquids. Volume 1. Fluid Mechanics*, 2nd ed.; John Wiley & Sons: New York, NY, USA, 1987.
5. Bonito, A.; Picasso, M.; Laso, M. Numerical simulation of 3D viscoelastic flows with free surfaces. *J. Comput. Phys.* **2006**, *215*, 691–716. [CrossRef]
6. Hirt, C.W. *Surface Tension Validation—Simple Test Problems*; Flow Science Report; Flow Science: Santa Fe, NM, USA, 2016; pp. 10–16.
7. Nguyen-Thien, T.; Tran-Cong, T.; Phan-Thien, N. An improved boundary element method for analysis of profile polymer extrusion. *Eng. Anal. Bound. Elem.* **1997**, *20*, 81–89. [CrossRef]
8. Spengler, G.; Wunsch, F. *Schmierung und Lagerung in der Feinwerktechnik*; VDI: Dusseldorf, Germany, 1970.
9. Lund, A.M.; Garcia, J.M.; Chambers, E. Line spread as a visual clinical tool for thickened liquids. *Am. J. Speech Lang Pathol.* **2013**, *22*, 566–571. [CrossRef]
10. Hilal, A.A.; Thom, N.H.; Dawson, A.R. The Use of Additives to Enhance Properties of Pre-Formed Foamed Concrete. *IACSIT Int. J. Eng. Technol.* **2015**, *7*, 286–293. [CrossRef]
11. Härth, M.; Schubert, D.W. Simple Approach for Spreading Dynamics of Polymeric Fluids. *Macromol. Chem. Phys.* **2012**, *213*, 654–665. [CrossRef]
12. de Gennes, P.G.; Brochard-Wyart, F.; Quéré, D. *Capillarity and Wetting Phenomena. Drops, Bubbles, Pearls, Waves*; Springer: New York, NY, USA, 2004.
13. Butt, H.-J.; Graf, K.; Kappl, M. *Physics and Chemistry of Interfaces*; Wiley: Weinheim, Germany, 2003.
14. Pérez, E.; Schäffer, E.; Steiner, U. Spreading Dynamics of Polydimethylsiloxane Drops: Crossover from Laplace to Van der Waals Spreading. *J. Colloid Interface Sci.* **2001**, *234*, 178–193. [CrossRef] [PubMed]
15. Fan, H. Liquid droplet spreading with line tension effect. *J. Phys. Condens. Matter* **2006**, *18*, 4481–4488. [CrossRef]
16. An, H.; Liu, G.; Craig, V.S. Wetting of nanophases: Nanobubbles, nanodroplets and micropancakes on hydrophobic surfaces. *Adv. Colloid Interface Sci.* **2015**, *222*, 9–17. [CrossRef] [PubMed]
17. Lee, F.; Chung, J.-E.; Kurisawa, M. An injectable enzymatically crosslinked hyaluronic acid-tyramine hydrogel system with independent tuning of mechanical strength and gelation rate. *Soft Matter* **2008**, *4*, 880–887. [CrossRef]
18. Winter, H.H.; Chambon, F. Analysis of Linear Viscoelasticity of a Crosslinking Polymer at the Gel Point. *J. Rheol.* **1986**, *30*, 367–382. [CrossRef]
19. Carre, A.; Gastel, J.-C.; Shanahan, M.E.R. Viscoelastic effects in the spreading of liquids. *Nature* **1996**, *379*, 432–434. [CrossRef]

fluids

MDPI

Article

Free-Flowing Shear-Thinning Liquid Film in Inclined μ-Channels

Angeliki T. Koupa, Yorgos G. Stergiou and Aikaterini A. Mouza *

Department of Chemical Engineering, Aristotle University of Thessaloniki, 54124 Thessaloniki, Greece; angelkoupa@auth.gr (A.T.K.); gstergiou@auth.gr (Y.G.S.)
* Correspondence: mouza@auth.gr; Tel.: +30-231-099-4161

Received: 10 December 2018; Accepted: 7 January 2019; Published: 10 January 2019

Abstract: Among the most important variables in the design of falling film microreactors (FFMRs) is the liquid film thickness as well as the gas/liquid interfacial area, which dictate the mass and heat transfer rates. In a previous work conducted in our lab the characteristics of a free-falling Newtonian liquid film have been studied and appropriate correlations have been proposed. In this work the geometrical characteristics of a non-Newtonian shear thinning liquid, flowing in an inclined open microchannel, have been experimentally investigated and design correlations that can predict with reasonable accuracy the features of a FFMR have been proposed. The test section used was an open μ-channel with square cross section (W_O = 1200 μm) made of brass which can be set to various inclination angles. The liquid film characteristics were measured by a non-intrusive technique that is based on the features of a micro Particle Image Velocimetry (μ-PIV) system. Relevant computational fluid dynamics (CFD) simulations revealed that the volume average dynamic viscosity over the flow domain is practically the same as the corresponding asymptotic viscosity value, which can thus be used in the proposed design equations. Finally, a generalized algorithm for the design of FFMRs, containing non-Newtonian shear thinning liquids, is suggested.

Keywords: free-flowing film; FFMR; inclined μ-channel; non-Newtonian; shear thinning; μ-PIV; meniscus

1. Introduction

The term process intensification describes the methodology which aims to build smaller, more compact and cheaper systems. Microfluidics, i.e., the technology that concerns the manipulation of fluids at the submillimeter scale, has shown considerable promise for improving diagnostics and biology research [1]. Among other potential applications, microfluidics includes fluid handling and quantitative analysis in healthcare and veterinary medicine [2]. During the last decades, engineers have developed various micro-devices which have at least one characteristic dimension less than 1 mm and can be potentially used in biological and medical applications. These microdevices are also characterized by low consumption of reagents and multifunctionality, as various unit operations can be combined in a single piece of equipment.

Microreactors are among the widely studied microdevices, since they have significant advantages over conventional ones [3]. Falling film microreactors (FFMRs) are liquid–gas phase contact devices in which the aim is to generate thin liquid films and to achieve large contact areas between the two phases. The large gas/liquid interface which provides enhanced mass transfer rates between the phases is included between the main advantages of micro-reactors [4]. Biomedical applications involve body fluids, such as blood, saliva, semen, mucus, that are fluids which exhibit non-Newtonian shear thinning behavior.

Most of the published work concerning FFMR deals with the conversion rate of specific reactions (e.g., [4,5]), while limited work [6,7] has been published on the characteristics of the interface in such devices. For example, Ishikawa et al. [8], have studied the CO_2 absorption into a wavy falling film of NaOH aqueous solution on a micro-baffled plate using two methods for visualizing the liquid film thickness. The first method uses the light reflection from a small angle (5°), that illuminates the liquid surface of a liquid film at another angle (45°). The second one is a particle injection method. More specifically the particles are PMMA slurry that can be detected and form the liquid film layer. Lokhat et al. [9], also studied, via CO_2 absorption experiments, the influence of a reaction plate orientation and gas flow rate on liquid phase mass transfer coefficient. They proposed correlations that are based on Nusselt's condensation theory. Tourvieille et al. [6], have visualized the liquid film thickness using fluorescence confocal microscopy, and they proposed correlations for determining the mass transfer coefficient, using Nusselt and Kapitza numbers. Yang et al. [10], who have investigated the liquid film thickness and the shape of the interface on an open channel FFMR using stereo digital microscopy, also proposed an empirical correlation, that predicts their experimentally measured liquid film thickness with 7% deviation. Patel et al. [11], have proposed another technique for the characterization of the interface between the liquid and the gas phase in a microchannel, using tracing particles and a microscope, and proved that this technique can reach the accuracy of 1.06 μm, without proposing any correlations for this specific method. Yu et al. [7], have experimentally measured the meniscus shape and the characteristics of the flow in inclined open rectangular microgrooves heat sinks using micro Particle Image Velocimetry (μ-PIV), and found that the meniscus shape is a parabolic arc.

In a previous work conducted in our lab the characteristics of a free falling Newtonian liquid film have been studied [12] and correlations have been proposed [13]. The aim of the present study is to extend our previous work [12,13] by performing experiments with non-Newtonian shear-thinning liquids and to check whether these correlations are valid for this type of fluids. Our ultimate goal is to propose an algorithm for the design of FFMRs.

2. Experimental Procedure

In a previous work of our group [12], a μ-PIV system, i.e., a non-intrusive method intended for measuring 2D velocity fields in microfluidics, was suitably adapted to measure the liquid film thickness and to reconstruct the shape of the interface of a free-flowing Newtonian liquid layer in an open μ-channel. In the proposed technique the area covered by the liquid phase is identified by recording the fluorescent particles used in μ-PIV for measuring the velocity field.

2.1. Experimental Setup

The experimental setup (Figure 1) comprises the test section, i.e., the open microchannel, which is connected to a syringe pump (AL-2000, World Precision Instruments®, Sarasota, FL, USA) to feed the liquid and the μ-PIV system. The μ-PIV system used consists of a high sensitivity charge-coupled device (CCD) camera (Hisense MkII, Dantec Dynamics®, Skovlunde, Denmark), which is connected to a microscope (Nikon Eclipse LV150, Nikon Corporation®, Tokyo, Japan), while the acquired images were processed by the Flow Manager Software (Dantec Dynamics, v4.00). Prior to measurements the fluids were traced by adding Nile red fluorescent carboxylate microspheres (Invitrogen, Carlsbad, MA, USA) with mean diameter of 1 μm. The measurements were conducted 30 diameters downstream from the inlet of the microchannel, where fully developed flow is established. The test section was placed on the microscope stage, which can be moved along its vertical axis with ten-micron accuracy. To obtain sufficiently magnified images a 20× air immersion objective with numerical aperture (NA) equal to 0.45 was used, which results in 3 μm depth of field. The maximum recording frame rate of the camera is 12.2 fps and that restricts the maximum frequency of the measurements to 12 Hz. The field of view was covering the whole channel width.

The test section used (Figure 2) was an open μ-channel, with square cross section (W_O = 1200 μm), made of brass. The test section was constructed by employing ultrahigh precision micromachining techniques, which have the unique advantage of being able to manufacture geometrically complex miniature components to high accuracy, with fine surface finish, in a wide range of engineering materials. The test-section used in the present was constructed using a five-axis ultra-precision micro milling machine that ensures both minimal surface roughness and straightness of the channels. Typical overall values of repeatability, surface finish and straightness achievable by this procedure are in the range of 1.0 μm, 0.040 μm Ra, and 0.2 μm over 100 mm of travel respectively. The channel was milled using a 500 μm cutter at 200 μm steps, while to ensure minimal surface roughness, a final 20 μm cut was performed.

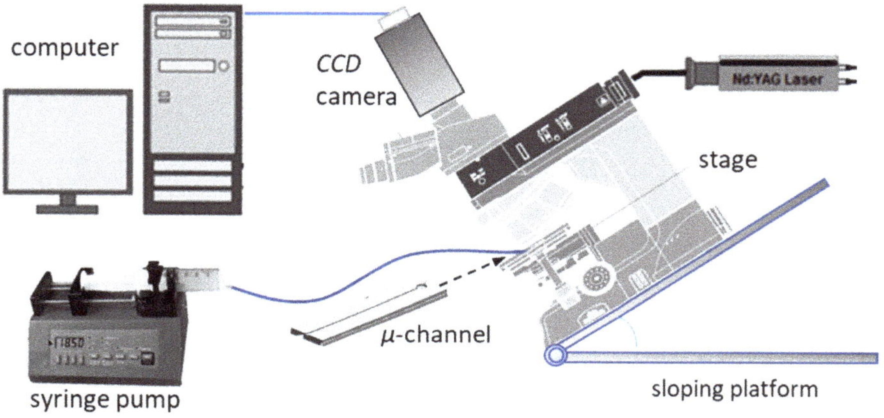

Figure 1. The experimental setup.

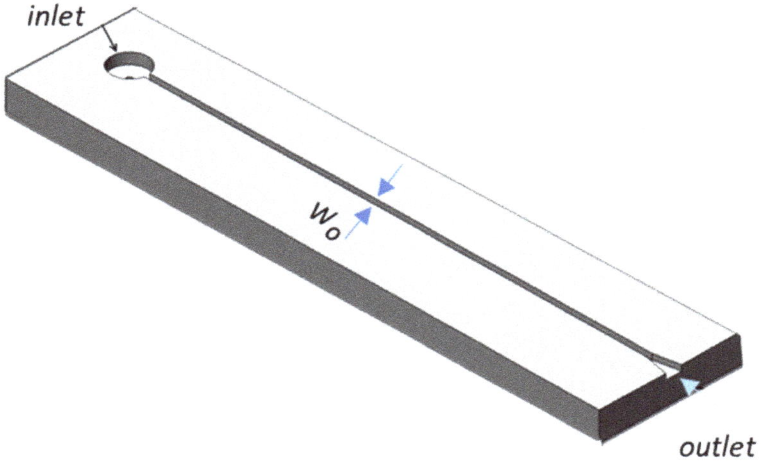

Figure 2. Details of the brass μ-channel.

To assure continuous free flow, the liquid phase enters by overflowing a circular region preceding the inlet of the microchannel. To conduct the measurements, the lens axis must be perpendicular to the test section and so the microscope was placed on platform that can be tilted up to 45° from the horizontal position (Figure 1). The measurements were conducted at three inclination angles, namely 20°, 25°, and 30°.

Two non-Newtonian shear thinning fluids were used. Presented in Table 1 are their physical properties, i.e., density, ρ, surface tension, σ, and the contact angle (for brass) measured by a tensiometer (KSV Cam200, KSV Instruments®, Helsinki, Finland), and dynamic viscosity, μ, measured by a magnetic rheometer (AR-G2, TA Instruments, Sussex, UK). All properties were evaluated at room temperature (20–22 °C), and the experiments were conducted under the same conditions. The viscosity of the shear-thinning non-Newtonian fluids used can be accurately described by the Herschel–Bulkley model [14]:

$$\mu = \frac{\tau_Y}{\dot{\gamma}} + K(\dot{\gamma})^{n-1} \quad (1)$$

where, τ_Y, is the yield stress, K, the viscosity consistency and n, the power law index. The physical properties of the fluids are shown in Table 1, while, for the liquids employed, Figure 3 presents the dependence of viscosity on shear rate.

Table 1. Physical properties of liquids used.

Index	Liquid	Refractive Index -	Contact Angle (°)	Density (kg/m³)	Surface Tension (mN/m)	Viscosity (Pa·s)
nW	100% water + 0.03% xanthan gum	1.340	74	998	72.1	$\mu = 0.003698\gamma^{-1} + 0.004339\gamma^{-0.1819}$
$nG20$	75% water + 25% glycerol w/w + 0.03% xanthan gum	1.360	74	1059	66.7	$\mu = 0.002952\gamma^{-1} + 0.006295\gamma^{-0.1535}$

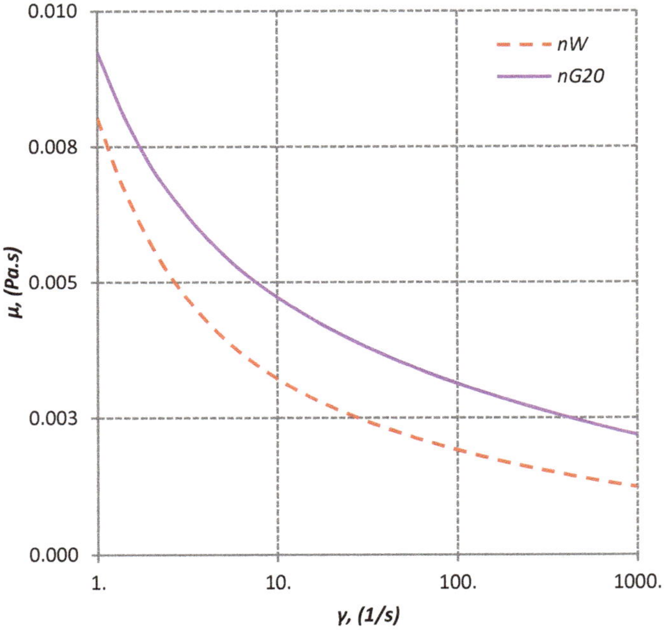

Figure 3. Viscosity curves of the two non-Newtonian liquids used.

2.2. Measuring Procedure

As already mentioned at the experimental set-up section, μ-PIV measurements require the injection of fluorescent tracing particles into the fluid. Consequently, in each focusing plane, the liquid phase corresponds to areas where light spots are visible, the external dark areas are identified as the walls of the microchannel, whereas the inner dark area corresponds to the gas phase (Figure 4). The use of proper microscope objectives ensures a narrow depth of field, which in turn increases the measuring accuracy, since the focus plane can be moved with a steady step of 10–20 μm resulting in to the recording of 20–25 planes per measurement.

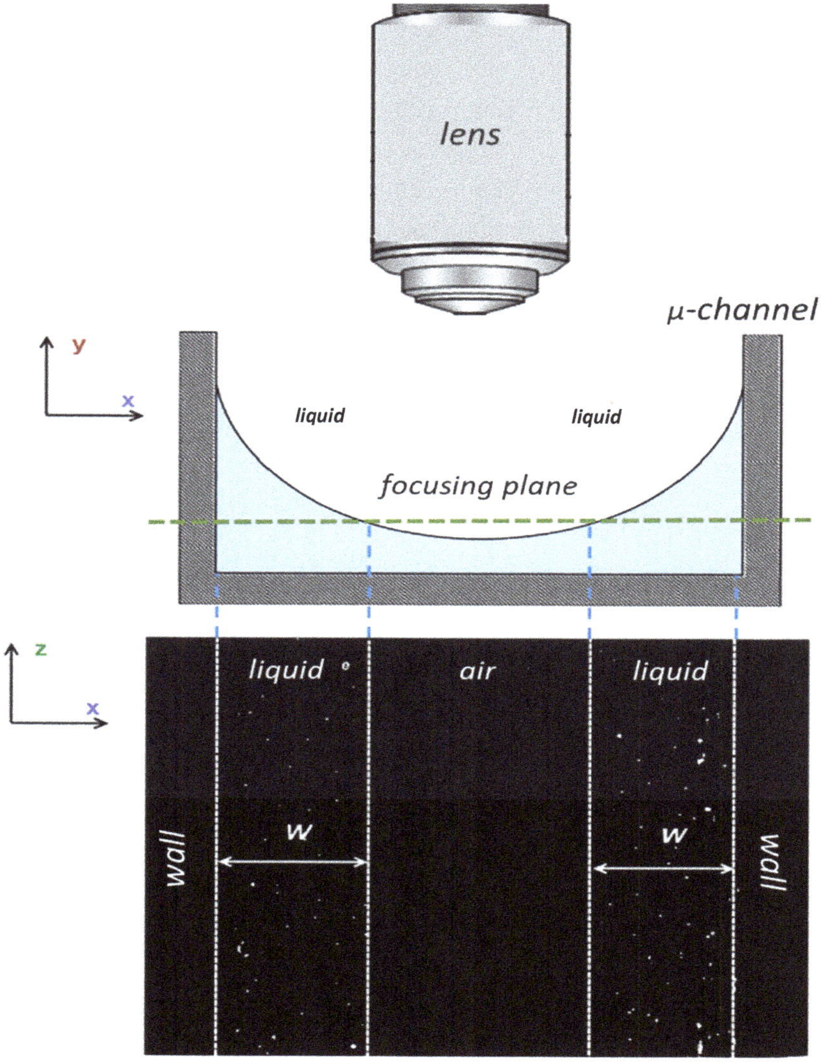

Figure 4. Identification of the liquid phase using fluorescent particles.

The measuring system was calibrated by measuring a known liquid film thickness, i.e., by measuring the thickness of the liquid layer obtained when a known liquid volume is placed in a vessel with known cross-section. It must be noted that during the measurements the displacement of the focusing plane due to refraction was also considered.

During each measurement, the plane, where the tracing particles are initially visible, is recognized as the first plane of the measurements. When the bottom of the meniscus, i.e., the plane where the liquid covers the entire area, is reached, no dark regions are observed. By further moving the focusing plane, an area where no particles are visible in reached, which is identified as the bottom of the microchannel. The liquid film thickness is the distance between the bottom of the meniscus and the bottom of the microchannel. The shape of the meniscus can be reconstructed by combining the results corresponding to the various focusing planes. A typical reconstruction of the film is presented in

Figure 5. As expected, the film is symmetrical with respect to the center plane of the channel and thus, to minimize effort, the measurements were taken only at one half of the channel. Figure 6 shows the main geometrical characteristics of the liquid film and a typical photo of it.

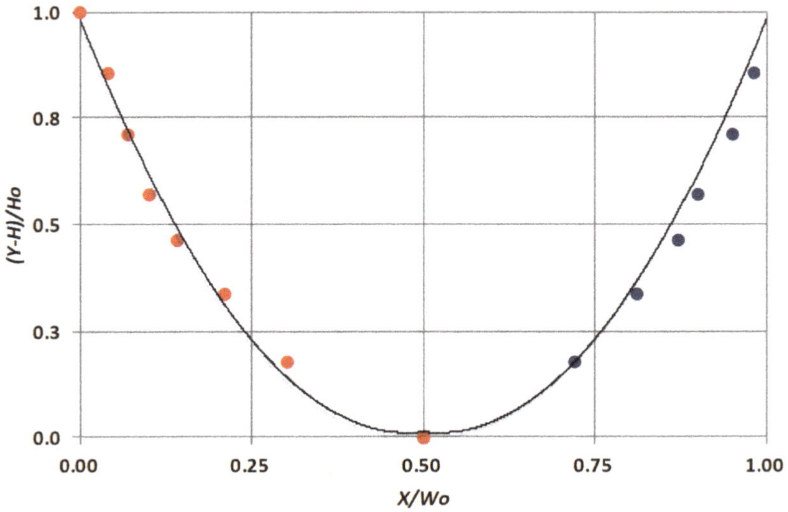

Figure 5. Typical measurements of the liquid film (nW, 40 mL/h, inclination angle 25°) where X is the distance from vertical wall and Y the distance from the channel bottom.

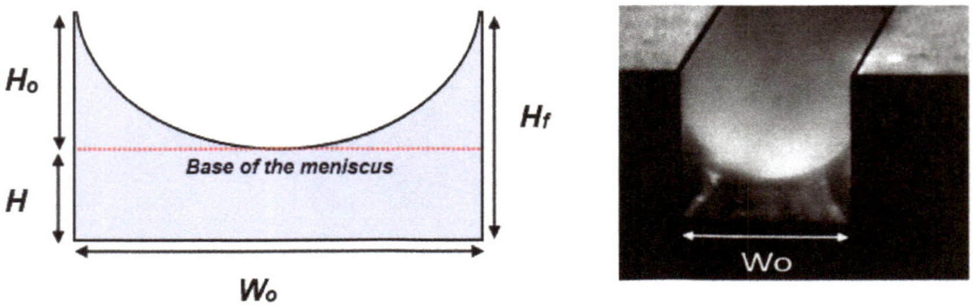

Figure 6. (**a**) Geometrical characteristics of the liquid film; (**b**) Typical photo of the film at the channel exit.

3. Results and Discussion

3.1. Liquid Film Thickness Calculation

In our previous work [13], according to the Rayleigh method of dimensional analysis, it was found that the film thickness H can be calculated as a function of modified dimensionless numbers, i.e., Reynolds (Re), Froude (Fr) and Capillary (Ca) number.

$$H/W_O = C\left(Re^a Ca^b Fr^d\right)^f \quad (2)$$

The dimensionless numbers use the channel width (W_O) as the characteristic length and the superficial velocity (U_S) defined as the volumetric flow rate divided by cross section of the meniscus. By inserting the aforementioned quantities, the dimensionless numbers are defined as follows:

$$Re = U_s W_O \rho/\mu = Q\rho/(\mu h) \quad (3)$$

$$Fr = U_s^2/g_\varphi W_O = Q^2/(g_\varphi W_O{}^3 h^2) \tag{4}$$

with g_φ being the component of the acceleration of gravity that acts in the direction of flow, i.e., $g_\varphi = g \sin(\varphi)$ and φ is the inclination angle from the horizontal.

$$Ca = \mu U_s/\sigma = \mu Q/W_o h\sigma \tag{5}$$

where ρ, μ and σ are the density, the viscosity and the surface tension of the fluid. In our case, where non-Newtonian liquids are used, an effective viscosity, i.e., a representative viscosity value along the whole domain, must be specified. To evaluate this viscosity, we have performed relevant CFD simulations.

3.2. Effective Viscosity Prediction

For the sake of computational simplicity, simulations were performed by forcibly imposing on the computational model the geometrical features of the liquid falling film, which are experimentally obtained. Namely, the experimentally measured parabolic surface profile was manually modeled and set as a free-slip wall boundary condition. At the bottom and at the two side walls of the domain the no-slip wall boundary condition was applied (Figure 7). Two liquid flow rate values were tested, i.e., 40 mL/h and 80 mL/h.

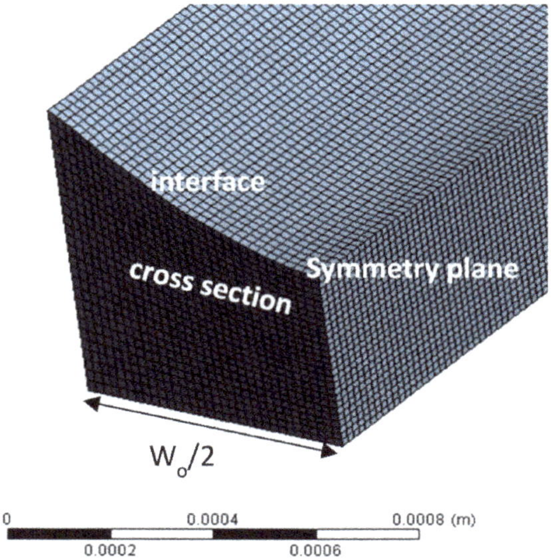

Figure 7. Grid density of the simulation.

The governing equations for the fluid flow are the Navier–Stokes equations and the continuity equation for incompressible and non-Newtonian flow. The fluid domain was discretized using tetrahedral elements and a symmetry plane was used to halve the problem for lowering computational costs. Total cell number varied for every case tested; from about 180,000 to about 240,000 cells. All simulations were performed using a commercial CFD code, the ANSYS CFX® (v. 19, ANSYS Inc., Canonsburg, PA, USA). ANSYS CFX® provides a finite volume method, a fully coupled solver for pressure and velocity coupling. The Direct Numerical Simulation (DNS) method for laminar flow was employed for the solution, as the flow in the μ-channel does not present any turbulent phenomena. Figure 8, shows the dynamic viscosity distribution at cross section, which was placed 100 widths downstream from the inlet to minimize inlet disturbances.

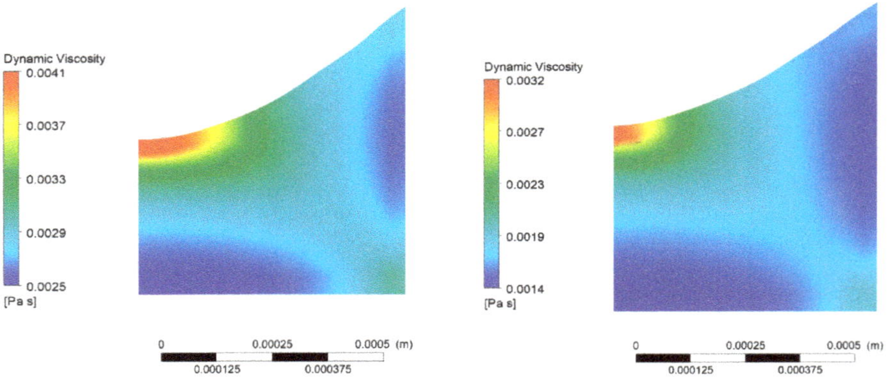

Figure 8. Dynamic viscosity distribution on a cross section located 30 W_O downstream from the channel entrance (**left to right**): *nW*_40; *nG*20_40.

Figure 8 presents typical dynamic viscosity distributions, calculated by CFD simulation, at a plane perpendicular to the flow direction and located 30 hydraulic diameters downstream from the liquid entrance. It worth mentioning that there are areas near the interface where the viscosity attains its higher value.

The volume averaged viscosity values predicted by the CFD simulation were compared with the asymptotic viscosity of the liquid, proving that the maximum difference is less than 8% (Table 2). We must also consider that the viscosity measurement has an inherent uncertainty of 8%. Consequently, the asymptotic viscosity, which is known a priori, can be used in Equations (3)–(5).

Table 2. Average and asymptotic viscosity for the liquids tested.

Fluid	μ_∞ (Pa·s)	μ_{ave} (Pa·s)	% Difference
*nG*20_80	0.0025	0.00260	4.0
*nG*20_40	0.0025	0.00270	8.0
*nW*_80	0.0014	0.00145	4.0
*nW*_40	0.0014	0.00150	7.1

The acquired data were appropriately fitted and the constants of Equation (2) were estimated (Table 3) and compared with the ones found for the same substrate type (i.e., brass) but for Newtonian fluids. It has been reported that for Newtonian fluids the constant *C* of Equation (2) depends on the type of the substrate material [15]. The present work proves that it also depends of the nature of the fluid, i.e., Newtonian or non-Newtonian, because for the same substrate type the non-Newtonian fluids result in thicker films. Although the volume averaged viscosity over the domain is almost the same as the corresponding asymptotic viscosity (Table 2), there are areas near the interface, where the viscosity values are much higher (Figure 8). Consequently, the liquid flow is expected to decelerate resulting in thicker films. As it can be seen in Figure 9, Equation (2) can predict with reasonable accuracy (less than ±15%) the liquid film thickness for the fluids, flow rates and inclination angles tested.

Table 3. Constants of Equation (2) for film thickness correlation.

Constants for Equation (1)	Present Work	Previous Work [13]
a	0.50	0.50
b	0.01	0.01
c	3.90	2.04
d	−0.56	−0.56
f	−0.86	−0.86

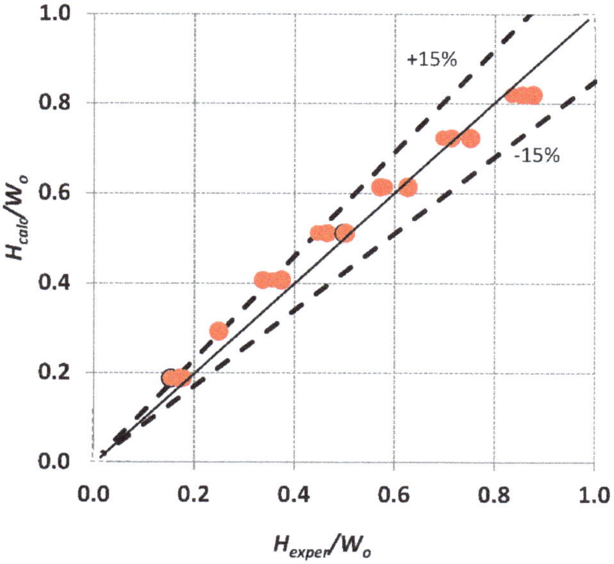

Figure 9. Comparison between experimental and calculated by Equation (2) liquid film thickness, H, normalized with respect to the channel width for the various non-Newtonian liquids, liquid flow rates and inclination angles tested.

3.3. Meniscus Shape

It is obvious that in *FFMRs* the concave shape of the interface augments the interfacial area. Thus, a characteristic that we must be able to determine during the design of an *FFMRs* is the shape and length of the gas–liquid interface (Figure 6a). In our previous work [13] it was shown that the shape of the meniscus can be accurately represented by a second order polynomial:

$$Y/\left(H_f - H\right) = C + A(X/W_o) + B(X/W_o)^2 \quad (6)$$

where X is the distance from the side wall, while the constants of the equation are defined as:

$$C = H_f - H \quad (7)$$

$$B = 4\left(H_f - H\right)/{W_o}^2 \quad (8)$$

$$A = -BW_O \quad (9)$$

where, H_f is the height where the three-phase contact is pinned, which for shallow μ-channels equals the channel's depth and H is the height of the meniscus (Figure 6a) that can be calculated by Equation (2).

Figure 10 presents the variation the meniscus height over the substrate divided by its maximum height along the channel width for the liquids, flow rates, and inclination angles studied. Figure 11 presents a comparison between the experimentally obtained meniscus profile and the one calculated by Equation (6). It is proved that Equation (6) can estimate the meniscus shape with reasonable accuracy (less than ±10%), for all the cases studied.

Figure 12 verifies the assumption that the asymptotic viscosity (μ_∞) can be used for calculating the liquid layer characteristics in place of volume averaged viscosity (μ_{ave}), estimated by CFD. More precisely, the shape of the meniscus calculated by Equations (2)–(7) using either μ_∞ or μ_{ave}, is compared with the one defined experimentally. Although the curve corresponding to μ_{ave} fits the experimental data better, the curve calculated using μ_∞, which is a priori known, can also predict the meniscus shape with reasonable accuracy, i.e., ±8%, compared with the one based on the experimental data.

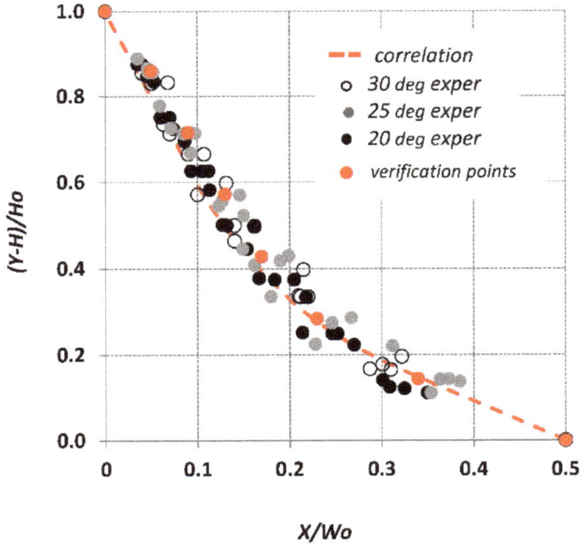

Figure 10. Comparison of the shape of the interface calculated by Equation (6) with relevant experimental data.

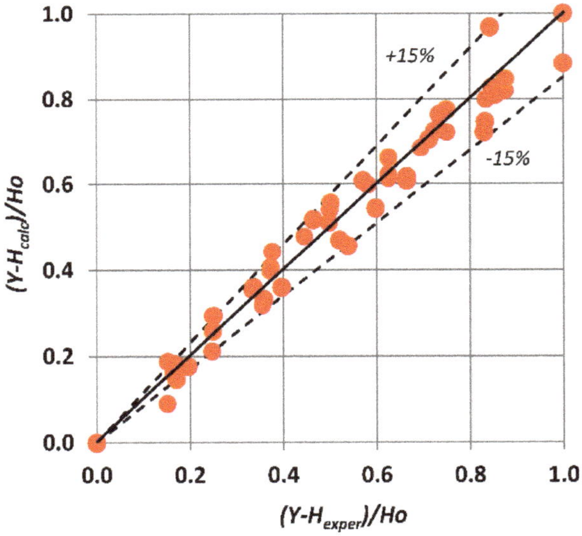

Figure 11. Comparison between experimental and estimated local meniscus height.

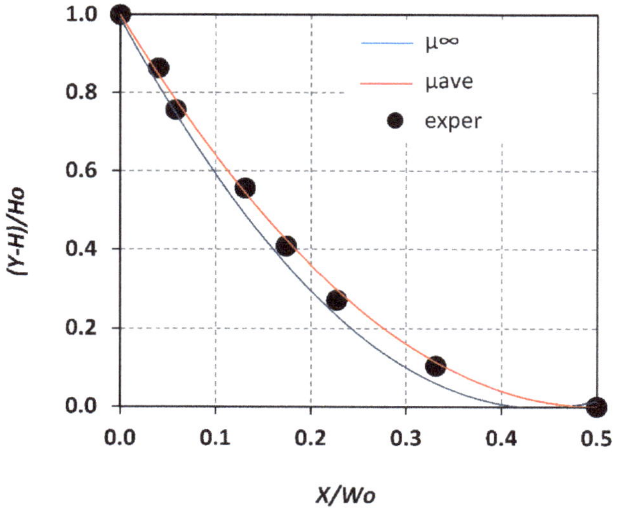

Figure 12. Comparison of experimental data with curves calculated using asymptotic and effective viscosity (nW, 40 mL/h, 25°).

The length of the parabolic arc (L), i.e., the length of the meniscus (Figure 6a), can be calculated with the following formula [16]:

$$L = \frac{1}{2}\sqrt{W_O{}^2 + 16H_O{}^2} + \frac{W_O{}^2}{8H_O} ln\left(\frac{4H_O + \sqrt{W_O{}^2 + 16H_O}}{Wo}\right) \tag{10}$$

where H_O and W_O are parameters defined in Figure 6a. The liquid phase cross section (A) was calculated by:

$$A = H_f W_O - \frac{2}{3} H_O W_O \tag{11}$$

Consequently, the specific surface (a) is the gas–liquid interfacial area (S) divided by the total fluid volume (V) in the reactor, that is:

$$a = \frac{S}{V} = \frac{L\cdot(reactorlength)}{A\cdot(reactorlength)} = \frac{L}{A} \tag{12}$$

whereas the fluid residence time per unit length of the reactor (τ) can be calculated by:

$$\tau = A/Q \tag{13}$$

Figures 13 and 14 present the effect of flow rate and inclination angle on the specific area and the residence time respectively. It is obvious that for a given flow rate both the specific area and the corresponding residence time can be controlled by altering the inclination angle.

Figure 15 presents a typical meniscus shape that corresponds to a flow rate of 40 mL/h and 25° inclination angle. The interfacial area calculated by the suggested procedure, i.e., Equations (6)–(11), is 15% lower than the one estimated using the experimental data. However, the deviation is considerably higher (i.e., around 30%), if a flat interface is assumed.

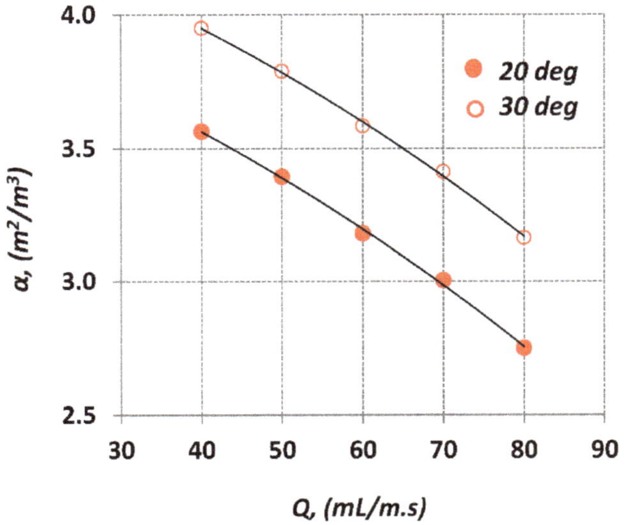

Figure 13. Effect of liquid flow rate and inclination on the specific surface, for *nW*.

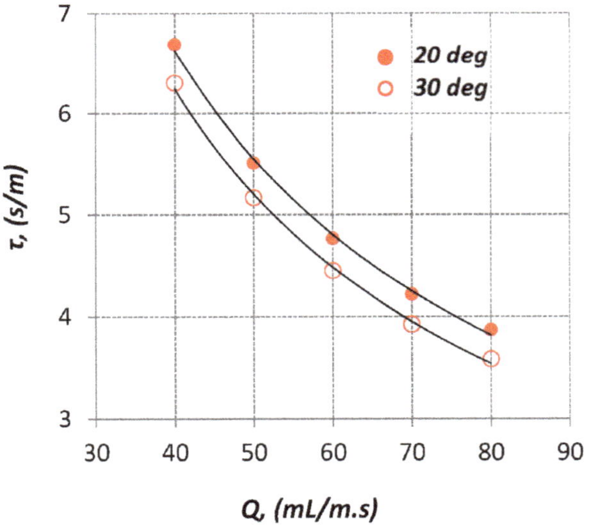

Figure 14. Effect of liquid flow rate and inclination on the residence time (for *nW*).

As a capstone of this study, a generalized algorithm for the design of FFMRs, is proposed (Figure 16). The input parameters of the algorithm are the physical properties and the flow rate of the liquid phase as well as the geometrical characteristics and the inclination angle of the μ-channel. Initially, the film thickness (H) is calculated by Equation (2). Subsequently, the shape of the interface can be determined by Equation (6) using the calculated film thickness (H) and the characteristics of the μ-channel. Finally, based on the calculated shape of the meniscus, the liquid phase cross section (A) can be calculated. Derived from this, important reactor design parameters, i.e., the specific surface (α) and the residence time (τ), can be specified.

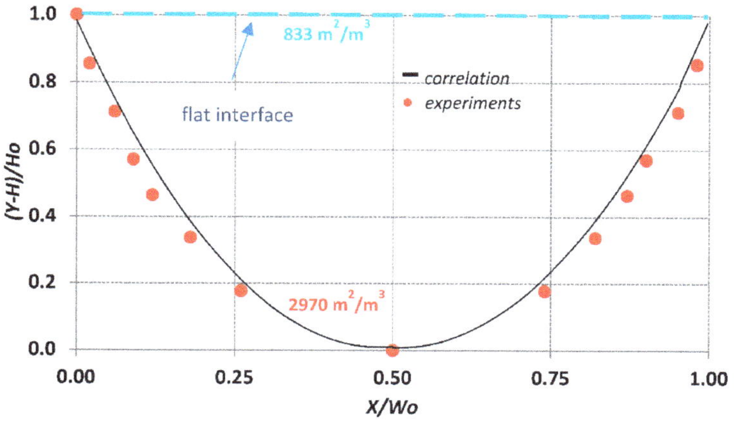

Figure 15. Comparison of specific surface between the flat interface and the meniscus shape calculated by the proposed correlation Equation (6) for nW, 40 mL/h, 25°.

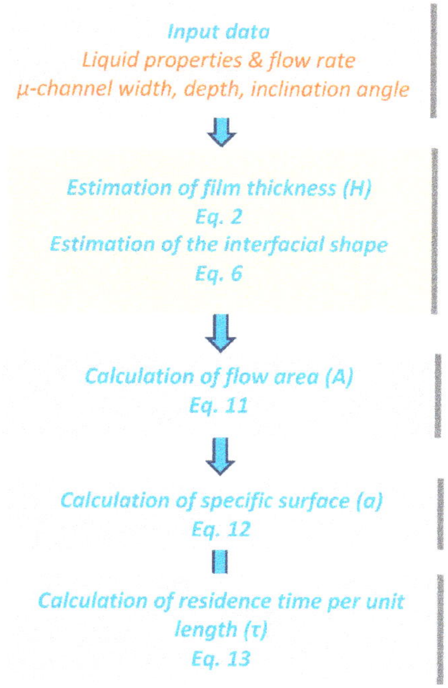

Figure 16. A general strategy for the design of falling film microreactors (FFMRs).

4. Conclusions

The most important features that must be considered during the design of FFMRs are the residence time of the liquid phase as well as the interfacial area per volume of fluid. Consequently, the thickness of the liquid film and the shape of the interface are key parameters for the design of such devices. In this work the geometrical characteristics of the liquid film, created when a non-Newtonian shear-thinning liquid flows in an inclined open microchannel made of brass, have been experimentally investigated.

Based on the experimental data, design correlations that can predict with reasonable accuracy the features of an FFMR design for biomedical applications have been proposed. The mean viscosity of the non-Newtonian fluid is calculated using relevant CFD simulations. It is also proved that the characteristics of the liquid film can be predicted with reasonable accuracy if the asymptotic viscosity (a priori known) is used in place of the calculated average one.

As a matter of fact, the design engineer could apply the proposed methodology to construct or rate an FFMR. Various input parameters, e.g., the inclination angle, can be changed and operating parameters (film thickness, interfacial area, residence time) can be recalculated by the same procedure. More work is certainly needed to investigate the dependence of the constant C (Equation (2)) on the channel material.

Author Contributions: A.A.M. had the initial conception of this work and designed the experiments; A.T.K. performed the experiments; Y.G.S. designed and performed the CFD simulations; A.A.M. interpreted the results; A.A.M., A.T.K. and Y.G.S. drafted the manuscript; A.A.M. reviewed and edited the manuscript.

Funding: This research received no external funding.

Acknowledgments: The authors would like to thank Spiros V. Paras for his helpful comments.

Conflicts of Interest: All authors state that there is no conflict of interest.

Nomenclature

A	liquid phase cross section	μm^2
Ca	capillary number	Dimensionless
Fr	Froude number	Dimensionless
g	acceleration of gravity	m/s^2
H_f	height of the microchannel	μm
H_o	height of the meniscus	μm
H	liquid film thickness	μm
L	length of the interface (meniscus)	μm
M	objective lenses magnitude	Dimensionless
m	mass flow rate	kg/s
NA	numerical aperture	Dimensionless
Q	volumetric flow rate	mL/h
Re	Reynolds number	Dimensionless
S	gas–liquid interfacial area	μm^2
V	total fluid volume	μm^3
X	distance from vertical wall	μm
Y	distance from the channel bottom	μm
W_O	width of microchannel	μm
Greek symbols		
α	specific surface	m^2/m^3
μ	liquid viscosity	Pa·s
σ	surface tension	N/m
τ	residence time/channel length	s/m
ρ	liquid density	kg/m^3
φ	inclination angle	°

References

1. Sackmann, E.K.; Fulton, A.L.; Beebe, D.J. The present and future role of microfluidics in biomedical research. *Nature* **2014**, *507*, 181–189. [CrossRef] [PubMed]
2. Yetisen, A.K.; Akram, M.S.; Lowe, C.R. Paper-based microfluidic point-of-care diagnostic devices. *Lab Chip* **2013**, *13*, 2210–2251. [CrossRef] [PubMed]
3. Hessel, V.; Löwe, H.; Schönfeld, F. Micromixers—A review on passive and active mixing principles. *Chem. Eng. Sci.* **2005**, *60*, 2479–2501. [CrossRef]

4. Chambers, R.D.; Holling, D.; Spink, R.C.; Sandford, G. Elemental fluorine. Part 13. Gas-liquid thin film microreactors for selective direct fluorination. *Lab Chip* **2001**, *1*, 132–137. [CrossRef] [PubMed]
5. Stavarek, P.; Le Doan, T.V.; Loeb, P.; De Bellefon, C. Flow visualization and mass transfer characterization of falling film reactor. In Proceedings of the 8th World Congress of Chemical Engineering, Montreal, QC, Canada, 23–27 August 2009.
6. Tourvieille, J.-N.; Bornette, F.; Philippe, R.; Vandenberghe, Q.; de Bellefon, C. Mass transfer characterisation of a microstructured falling film at pilot scale. *Chem. Eng. J.* **2013**, *227*, 182–190. [CrossRef]
7. Yu, D.; Hu, X.; Guo, C.; Wang, T.; Xu, X.; Tang, D.; Nie, X.; Hu, L.; Gao, F.; Zhao, T. Investigation on meniscus shape and flow characteristics in open rectangular microgrooves heat sinks with micro-PIV. *Appl. Therm. Eng.* **2013**, *61*, 716–727. [CrossRef]
8. Ishikawa, H.; Ookawara, S.; Yoshikawa, S. A study of wavy falling film flow on micro-baffled plate. *Chem. Eng. Sci.* **2016**, *149*, 104–116. [CrossRef]
9. Lokhat, D.; Domah, A.K.; Padayachee, K.; Baboolal, A.; Ramjugernath, D. Gas–liquid mass transfer in a falling film microreactor: Effect of reactor orientation on liquid-side mass transfer coefficient. *Chem. Eng. Sci.* **2016**, *155*, 38–44. [CrossRef]
10. Yang, Y.; Zhang, T.; Wang, D.; Tang, S. Investigation of the liquid film thickness in an open-channel falling film micro-reactor by a stereo digital microscopy. *J. Taiwan Inst. Chem. Eng.* **2018**. [CrossRef]
11. Patel, R.S.; Garimella, S.V. Technique for quantitative mapping of three-dimensional liquid–gas phase boundaries in microchannel flows. *Int. J. Multiph. Flow* **2014**, *62*, 45–51. [CrossRef]
12. Anastasiou, A.D.; Makatsoris, C.; Gavriilidis, A.; Mouza, A.A. Application of μ-PIV for investigating liquid film characteristics in an open inclined microchannel. *Exp. Therm. Fluid Sci.* **2013**, *44*, 90–99. [CrossRef]
13. Anastasiou, A.D.; Gavriilidis, A.; Mouza, A.A. Study of the hydrodynamic characteristics of a free flowing liquid film in open inclined microchannels. *Chem. Eng. Sci.* **2013**, *101*, 744–754. [CrossRef]
14. Chhabra, R.P.; Richardson, J.F. *Non-Newtonian Flow and Applied Rheology: Engineering Applications*; Butterworth-Heinemann: Oxford, UK, 2011; ISBN 978-0-08-095160-7.
15. Anastasiou, A.D.; Al-Rifai, N.; Gavriilidis, A.; Mouza, A.A. Prediction of the characteristics of the liquid film in open inclined micro-channels. In Proceedings of the 4th Micro and Nano Flow Conference 2014, London, UK, 6–10 September 2014.
16. Spiegel, M.R.; Liu, J. *Mathematical Handbook of Formulas and Tables*; McGraw-Hill: New York, NY, USA, 1999; ISBN 978-0-07-038203-9.

fluids

MDPI

Article

A Simplified Model for Predicting Friction Factors of Laminar Blood Flow in Small-Caliber Vessels

Aikaterini A. Mouza [1,*], Olga D. Skordia [1], Ioannis D. Tzouganatos [2] and Spiros V. Paras [1]

1 Department of Chemical Engineering, Aristotle University of Thessaloniki, 541 24 Thessaloniki, Greece; skordiao@auth.gr (O.D.S.); paras@auth.gr (S.V.P.)
2 Department of Chemical Engineering, Imperial College, London SW7 2AZ, UK; ioannis.tzouganatos14@imperial.ac.uk
* Correspondence: mouza@auth.gr; Tel.: +30-231-099-4161

Received: 2 September 2018; Accepted: 16 October 2018; Published: 19 October 2018

Abstract: The aim of this study was to provide scientists with a straightforward correlation that can be applied to the prediction of the Fanning friction factor and consequently the pressure drop that arises during blood flow in small-caliber vessels. Due to the small diameter of the conduit, the Reynolds numbers are low and thus the flow is laminar. This study has been conducted using Computational Fluid Dynamics (CFD) simulations validated with relevant experimental data, acquired using an appropriate experimental setup. The experiments relate to the pressure drop measurement during the flow of a blood analogue that follows the Casson model, i.e., an aqueous Glycerol solution that contains a small amount of Xanthan gum and exhibits similar behavior to blood, in a smooth, stainless steel microtube (L = 50 mm and D = 400 μm). The interpretation of the resulting numerical data led to the proposal of a simplified model that incorporates the effect of the blood flow rate, the hematocrit value (35–55%) and the vessel diameter (300–1800 μm) and predicts, with better than ±10% accuracy, the Fanning friction factor and consequently the pressure drop during laminar blood flow in healthy small-caliber vessels.

Keywords: pressure drop; CFD; Casson fluid; blood; hematocrit; small vessel; microfluidics

1. Introduction

The flow behavior of non-Newtonian fluid in small-caliber tubes is of high interest in practical applications (e.g., flow, mixing and separation of various biological species in microchips) [1]. Blood exhibits non-Newtonian behavior, meaning that its viscosity depends on shear rate. The ability to predict blood flow pressure drop is essential in therapy strategy and the design of surgical repairments and implantable medical devices [2]. An investigation on blood flow in small arteries is of both fundamental interest and practical significance [2].

Blood is a two-phase suspension of plasma and cells that behaves as a non-Newtonian shear-thinning viscoplastic liquid. The plasma exhibits Newtonian behavior with a viscosity ranging from 1.10 to 1.35 mPa·s at 37 °C. It comprises of proteins (albumin, globulins and fibrinogen) that represent about 7–8 wt% of plasma, glucose, clotting factors, inorganic ions, dissolved gases, hormones and other substances at low concentration [3]. The most abundant cells are the red blood cells (RBCs) or erythrocytes comprising about 95% of the cellular component of blood while the remaining 5% are comprised of white cells and platelets. The non-Newtonian shear-thinning character of blood results from variations in the aggregation and deformation of the red cells. On the other hands, its viscosity is primarily affected by the volume percentage of red blood cells in blood, i.e., the hematocrit value.

The equations that define non-Newtonian flow, although well established, are highly complex compared to those defined for Newtonian flow [4]. This is why in many papers, blood is treated as a Newtonian fluid. Although this assumption holds practically true for high shear rates, i.e., greater than 1000 s^{-1}, it is not valid for blood flow in small vessels [5], where the Reynolds numbers are low, and the flow is laminar.

Research that addresses non-Newtonian fluid flow is based on the pioneering work of Metzner and Reed [6] in 1955. According to their approach the same friction factor chart could be used for Newtonian and time-independent non-Newtonian fluids in the laminar or turbulent region. However, only provided that an appropriate effective Reynolds number can be estimated. In this way the problem reduces to the quest of an appropriate effective viscosity, which would relate the behavior of the non-Newtonian fluid to an equivalent one of a hypothetical Newtonian fluid [7]. Nevertheless, the calculations seem to be complex and rather laborious because the suggested equations must be solved iteratively.

Many models have been developed to describe blood viscosity, e.g., Casson, Hershel-Bulkley, Carreau and Quemada [8]. Chilton and Stainsby [4] suggested a model for Hershel-Bulkley fluids that iteratively computes the friction factor both in laminar and turbulent flow in pipes. However, this method does not take into account the effect of hematocrit. Cruz et al. [9] proposed a generalized method for predicting friction factors in fully developed non-Newtonian laminar flow in circular pipes using several viscoelastic models i.e., Herschel-Bulkley, Bingham, Casson and Carreau-Yasuda. Their method, apart from the fluid velocity, the pipe diameter and the parameters of each rheological model utilized an apparent flow behavior index, where estimation was quite complicated. Therefore, the most prevalent model for predicting blood viscosity is the Casson model [10] because its constants can be expressed as a function of hematocrit [8] and it is reported to lead to reliable results [11].

The pressure drop exerted during blood flow depends mainly on the diameter of a blood vessel, the blood flow rate as well as the hematocrit value [11]. The aim of this study was to provide engineers and physicians with a simple, straightforward algorithm that could be applied to the prediction of the pressure losses, or equivalently, the Fanning friction factor during blood flow in small-diameter healthy vessels. As the experiments were difficult to perform and time consuming, the effect of the various parameters was evaluated by performing a series of Computational Fluid Dynamics (CFD) simulations, using a previously validated code. Due to the small diameter of the conduit, the corresponding Reynolds numbers were low (i.e., less than 50) and consequently the flow was regarded laminar.

2. Experimental Setup and Procedure

It is common practice prior to proceeding with simulations to validate the CFD code using data acquired by performing relevant experiments. The experimental setup used for the pressure drop measurements (Figure 1) comprised the test section (Figure 2), two syringe pumps (AL-2000, World Precision Instruments®, Sarasota, FL, USA), a three-way valve and a digital pressure transducer (68035, Cole Palmer, Vernon Hills, IL, USA).

The use of two pumps was necessary to cover the desired Reynolds number range (i.e., 1–50). The three-way valve allowed for the refilling of the syringes without exposing the system to the atmospheric air. The pressure was monitored by connecting one of the manometer tubes to the microtube inlet, while the second was left open to the atmosphere.

The test section (Figure 2) consisted of a 50 mm microtube, i.e., a stainless steel (SS 304) chromatography needle (Gauge 33, Hamilton, Merck, Darmstadt, Germany) adjusted on a polymethyl methacrylate (PMMA) cube and connected to the feed and the manometer. All experiments were conducted at room temperature (i.e., 20 ± 1 °C).

The functionality of the setup was confirmed by performing experiments with a Newtonian fluid (i.e., water), while the CFD model was validated with data obtained from performing experiments with a non-Newtonian fluid, i.e., a blood analogue.

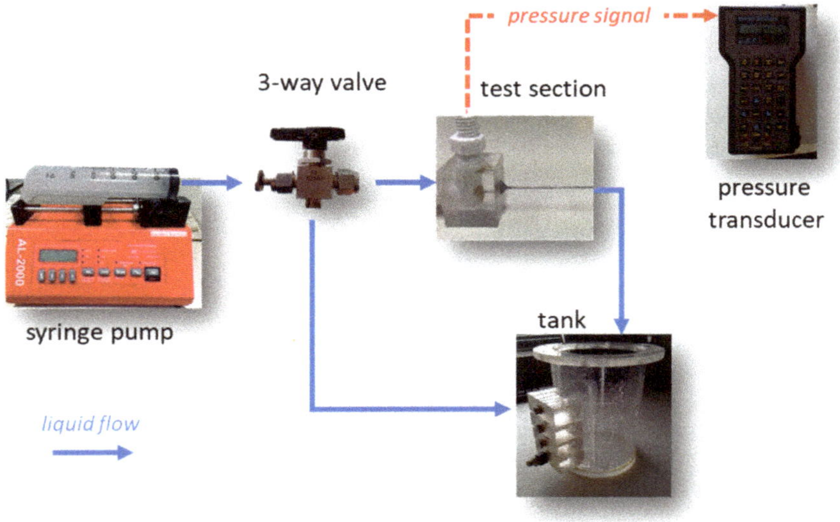

Figure 1. Experimental setup for pressure drop measurements.

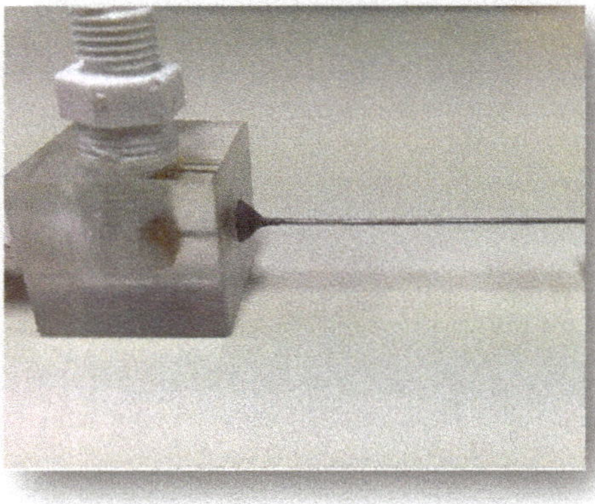

Figure 2. Detail of the pressure drop measuring test section.

Blood Analogue

As experiments with blood are difficult to perform due to coagulation, blood-mimicking fluids were used, i.e., fluids whose rheological properties are similar to blood and whose viscosity follow the Casson model. As previously mentioned, the constants of the Casson model depend on the hematocrit value where the normal range is between 30% and 55% [8].

The viscosity of blood can be expressed by Equation (1) [3]:

$$\mu = \left(\sqrt{\frac{\tau_y}{\gamma}} + \sqrt{n_N} \right)^2 \tag{1}$$

where μ is the viscosity of blood, γ is the shear rate, τ_y is the yield stress and n_N is the viscosity corresponding to high shear rates (asymptotic value). The yield stress is a measure of the amount of energy required to break down the aggregates of red blood cells formed at very low shear rates.

Merrill [8] extensively investigated blood rheology and confirmed the strong relationship between viscosity and hematocrit (H_t) and suggested that the terms n_N and τ_y of Equation (1) could be expressed as functions of hematocrit, i.e.,

$$n_N = n_p\left[1 + 0.025H_t + 7.35 \cdot 10^{-4}H_t^2\right] \quad (2)$$

where n_p is the viscosity of the plasma,

$$\tau_y = A(H_t - H_{tc})^3 \quad (3)$$

and H_{tc} is the critical hematocrit below which the yield stress (τ_y) can be considered negligible. For normal blood, H_{tc} ranges between 4 and 8 and A is a constant, ranging between 0.6×10^{-7} and 1.2×10^{-7} Pa. These expressions are employed hereafter for predicting blood viscosity as a function of hematocrit [8]. In this study, the values selected for A and H_{tc} were 0.9×10^{-7} Pa, and 6, respectively, i.e., the middle values of the corresponding range.

For the sake of simplicity, the variations in viscosity previous mentioned, could be attributed to several factors and were expressed in the simulations as a function of hematocrit alone. Blood density (ρ) was assumed to be constant, i.e., independent of the hematocrit value investigated and equal to 1060 kg/m^3.

Thus, human blood with a H_t of ~55% can be simulated by a 30% v/v aqueous glycerol solution that contains 0.035% w/v xanthan gum, i.e., a polysaccharide that acts as a rheology modifier and renders the fluid non-Newtonian. The viscosity curve of the fluid was measured in our laboratory via a magnetic rheometer (AR-G2, TA Instruments, Sussex, UK), for shear rates between 1–1000 s^{-1} (Figure 3) resulting in an excellent fit shown by a Casson-type curve (Equation (1)).

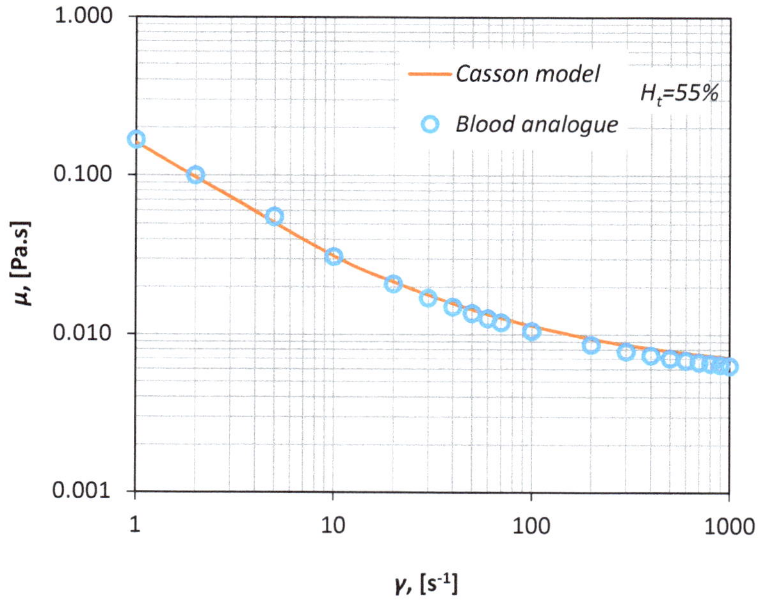

Figure 3. Viscosity measurements of the blood analogue (H_t = 55%).

3. Numerical Simulations

CFD modeling is considered a dependable tool for studying blood flow characteristics in small-caliber vessels [11]. In this study, a CFD code (ANSYS CFX 18.1, ANSYS, Inc., Karnosboro, PA,

USA) was employed for simulating the flow. The code comprised of the usual parts of a standard CFD code, i.e., the pre-processing part, where the computational domain and the grid are constructed, the solver, where the discretized differential equations are iteratively solved and the post processing part where the results are analyzed and displayed in an illustrative way.

A blood vessel was modeled as a three-dimensional (3D) computational domain, while the geometry of the computational domain and the mesh were designed using the parametric features of the ANSYS Workbench package (version 18.1). The length of the conduit was 5 cm, the fluid density was assumed to be constant (i.e., 1060 kg/m^3), while the diameter of the conduit was a parametric variable and was set to be between 0.3–1.8 mm. With the aim of reducing memory consumption and CPU time and as the geometry consisted of two symmetry planes, only one fourth of the domain was used (Figure 4).

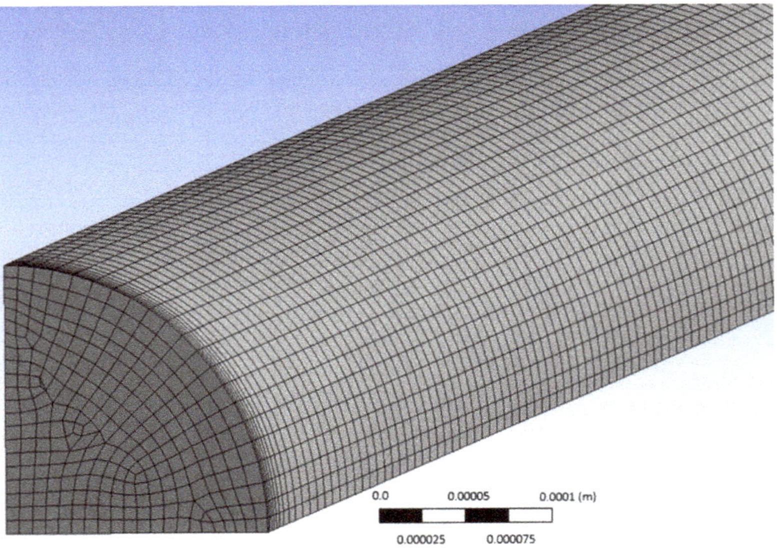

Figure 4. Typical computational domain of the simulations.

Due to the small characteristic dimension of the conduit, the flow is laminar. Hence, the Direct Numerical Simulation (DNS) model was selected while the high-resolution advection scheme was used for the discretization of the momentum equations. A custom-made high-performance unit for parallel computing was used (24 nodes, 64 GB RAM). The simulations were run in a steady state, the vessel walls were considered smooth and the non-slip boundary condition was imposed at the walls, while the flow rate was kept constant for each run.

As the numerical diffusion in the CFD calculations are known to influence the accuracy of the calculations, an optimum grid density was chosen by performing a grid dependency study. Figure 5 shows typical results that illustrate the dependence of pressure drop on the number of cells and corresponds to the maximum flow rate that was tested. Hence, 270,000 cells were chosen to ensure that the solution was independent of the grid density.

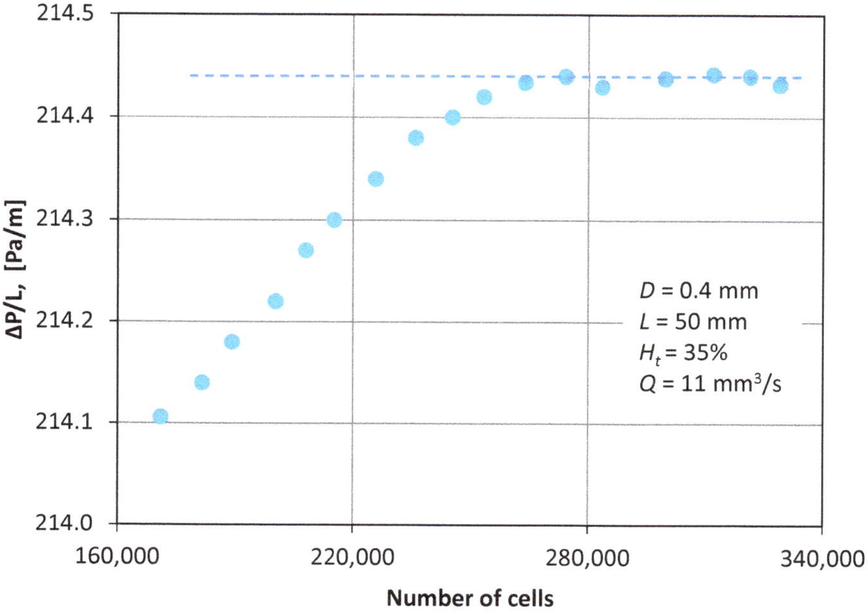

Figure 5. Grid dependency study (D = 0.4 mm, L = 50 mm, Q = 11 mm^3/s, H_t = 35%).

3.1. Code Validation

Usually, the validity of a CFD code is checked by comparing the numerical results with relevant experimental data. In the present study, appropriate experiments were performed using the experimental setup described in Section 2. Furthermore, the experimental data presented in Figure 6 together with the computational results proved to be in very good agreement (±10%). Since the viscosity of the non-Newtonian fluids was not constant, the pressure drop was plotted versus a modified Reynolds number Re^* (Figure 6)

$$Re^* = \frac{UD\rho}{\mu^*} \tag{4}$$

which in place of the fluid viscosity uses an effective viscosity μ^*, that corresponds to the pseudo-shear rate γ^* defined by Equation (5) [3].

$$\gamma^* = \frac{U}{D} = \frac{4Q}{\pi D^3} \tag{5}$$

Here, U is the average fluid velocity, Q is the volumetric blood flow rate and D is the inside diameter of the vessel.

Figure 7 presents typical CFD results of the pressure distribution along the axis of the conduit. The shape of the curve (i.e., straight line) denotes that the flow was fully developed.

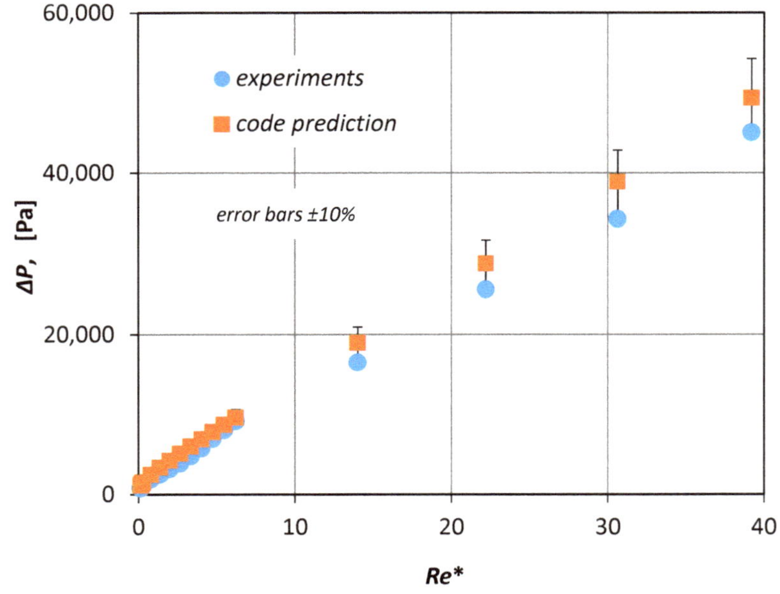

Figure 6. Validation of the Computational Fluid Dynamics (CFD) code (error bars refer to ±10%).

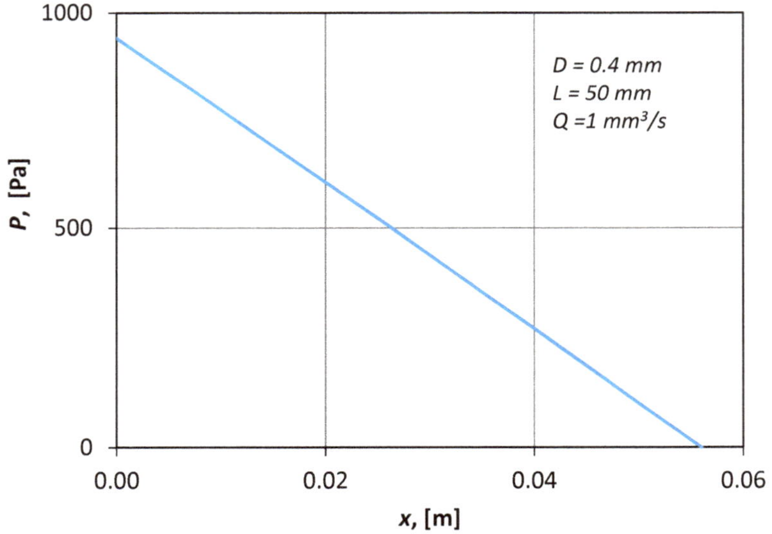

Figure 7. Typical simulation results of the pressure distribution across a small vessel (L = 50 mm, D = 0.4 mm and Q = 1 mm^3/s).

3.2. Numerical Procedure

A parametric study was performed by employing the Design Exploration features of the ANSYS Workbench® package. The design variables selected along with the imposed upper and lower bounds are presented in Table 1. The upper bound of the vessel inside diameter corresponds to the larger arterioles and venules of an adult male [12], while the corresponding lower bound was chosen to be

500 μm to avoid the consequences of the Fahraeus-Lindqvist effect [12]. Due to this effect in small vessels (smaller than 300 μm), red blood cells tend to drift towards the central axis of the vessel, forming a cell-free layer called a plasma layer along the vascular wall. This effect results in an apparent blood viscosity, which declines substantially with decreasing diameter [13]. The hematocrit range that was chosen is typical for healthy adult humans [8] and the blood flow rate bounds imposed are typical for such μ-vessels.

Table 1. Constraints of the design variables.

Parameter	Lower Bound	Upper Bound
Vessel inside diameter (mm)	0.50	1.80
Hematocrit (%)	35	50
Blood flow rate (mm^3/s)	7.0	88.0

To extract the necessary information from a limited number of test cases, the Design of Experiments (DOE) methodology was followed. In the present study, the effect of the design parameters was investigated by performing a series of "computational experiments" for certain values of the design parameters chosen by employing the Box-Behnken method [14], i.e., an established DOE technique.

A DOE technique leads to reliable conclusions from the least possible number of design points.

Table 2 presents the various design points selected. Based on the computational results, the aim is to formulate appropriate design correlations that can be applied for the prediction of friction factor values during laminar blood flow in micro vessels.

Table 2. Design and verification points.

Design Points						Verification Points		
Box-Behnken			Additional Points					
Q mm^3/s	D mm	H_t %	Q mm^3/s	D mm	H_t %	Q mm^3/s	D mm	H_t %
72.0	1.15	35	6.0	0.40	35	6.7	0.40	55
72.0	1.15	50	8.5	0.40	35	7.0	0.50	40
72.0	1.80	43	11.0	0.40	35	10.0	0.40	55
72.0	0.50	43	13.0	0.40	35	11.7	0.40	55
80.0	0.50	35	16.0	0.40	35	15.0	0.40	55
80.0	0.50	50	18.0	0.40	35	20.0	0.30	45
80.0	1.15	43	21.0	0.40	35	33.4	0.40	55
80.0	1.80	35	23.0	0.40	35	40.0	0.90	50
80.0	1.80	50	25.0	0.40	35	66.8	0.40	55
88.0	1.15	35	45.0	0.40	35	68.0	1.20	35
88.0	1.15	50	50.0	0.60	37	83.0	0.60	37
88.0	1.80	43	7.0	0.30	43	83.5	0.40	55
88.0	0.50	43	-	-	-	-	-	-

The first three columns of Table 2 include the design points dictated by the Box-Behnken method. The following three columns contain some additional points that extend the D range to 0.3 mm. The last three columns of Table 2 present appropriate verification points that lay within the range of design points. These were arbitrarily chosen and were used for testing the applicability of the design equation.

4. Results

In Figure 8, the measured pressure drop values for various blood flow rates (Q), are compared with pressure drop losses (Equation (6)):

$$\frac{\Delta P}{L} = f\frac{2}{D}\rho U^2 \quad (6)$$

where f is the Fanning friction factor given by:

$$f = {}^{16}/_{Re_\infty} \quad (7)$$

where Re_∞ is a Reynolds number that uses the asymptotic value (μ_∞) of blood viscosity. From Figure 8 it is obvious that by no means can the correlation for Newtonian fluids be applied to non-Newtonian fluids because it underestimates ΔP by 30%.

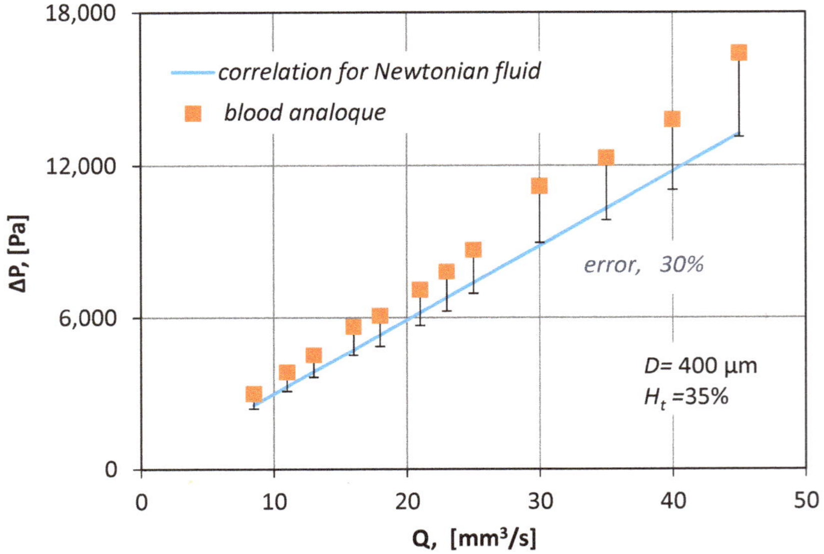

Figure 8. Comparison of experimental results for the blood analogue with the theoretical prediction of ΔP using the correlation for Newtonian fluids ($f = 16/Re_\infty$).

Initially, the numerically predicted pressure drop value was used to calculate Equation (6), the Fanning friction factor that corresponds to each "experiment". In Figure 9, the calculated friction factors were compared with the values predicted by Equation (8) that related the Fanning friction factor with Re^* (defined by Equation (4)), i.e.,

$$f = {}^{16}/_{Re^*} \quad (8)$$

The calculated friction factors deviated considerably from the ones predicted by Equation (8) especially for the lower Re^* values. It was obvious that a new correlation had to be formulated.

As blood is a non-Newtonian fluid, we attempted to strengthen the contribution of its viscoplastic nature [7] by multiplying Re^* with the Bingham number, which for viscoplastic materials, expresses the relative importance of yield stress to viscous stress. This is defined as:

$$Bm = {}^{\tau_y D}/_{\mu^* U} \quad (9)$$

It is found that the friction factor can be well predicted using Equation (10):

$$f = 5.974\ Bm^{-0.266}\ Re^{*-1.064} \quad (10)$$

Coefficients are determined by fitting the numerical data that correspond to the design points (Figure 10). The validity of Equation (10) was further tested by comparing it with the data obtained using the verification points. Figure 10 shows that the proposed correlation was in excellent agreement with all the results.

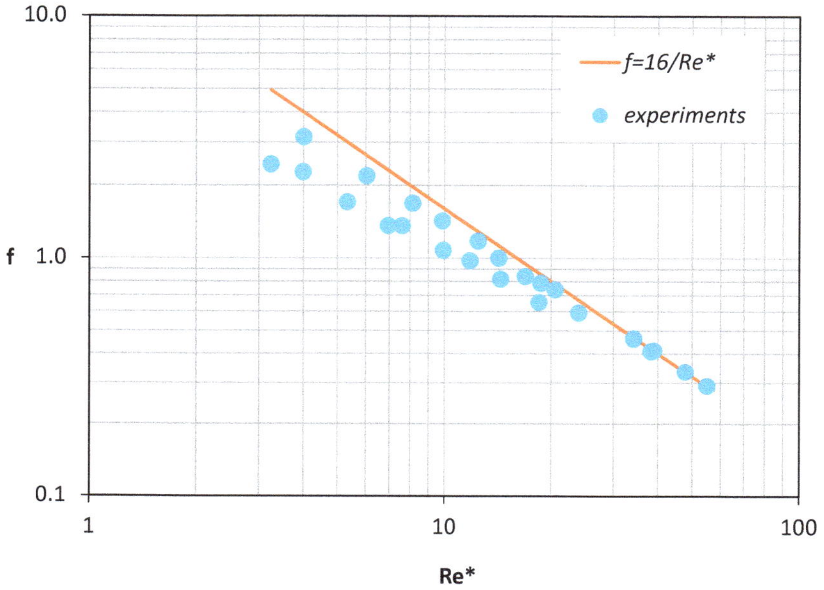

Figure 9. Friction factor versus *Re**.

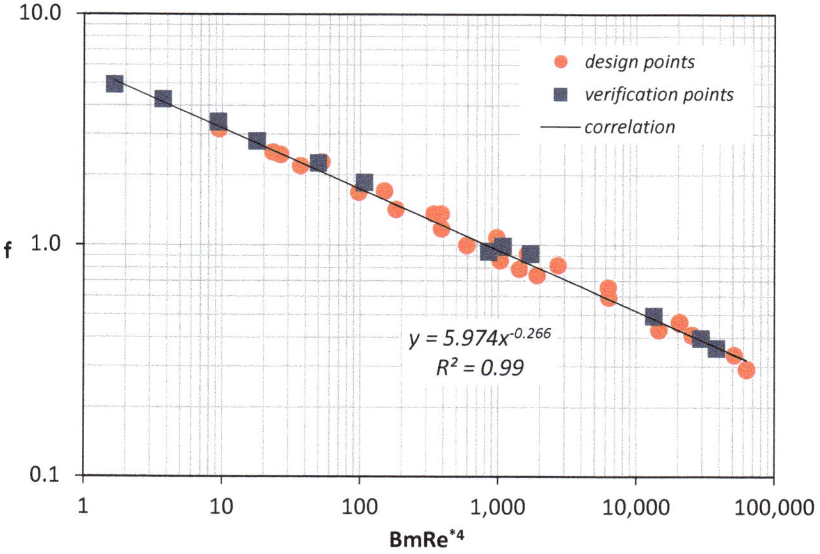

Figure 10. Friction factor versus the dimensionless group $BmRe^{*4}$.

Figure 11 compares the friction factor values (f_{calc}), calculated by Equation (10) with the values that resulted from the CFD simulations (f_{CFD}). It was proven that Equation (10) could predict the Fanning friction factor with 10% uncertainty.

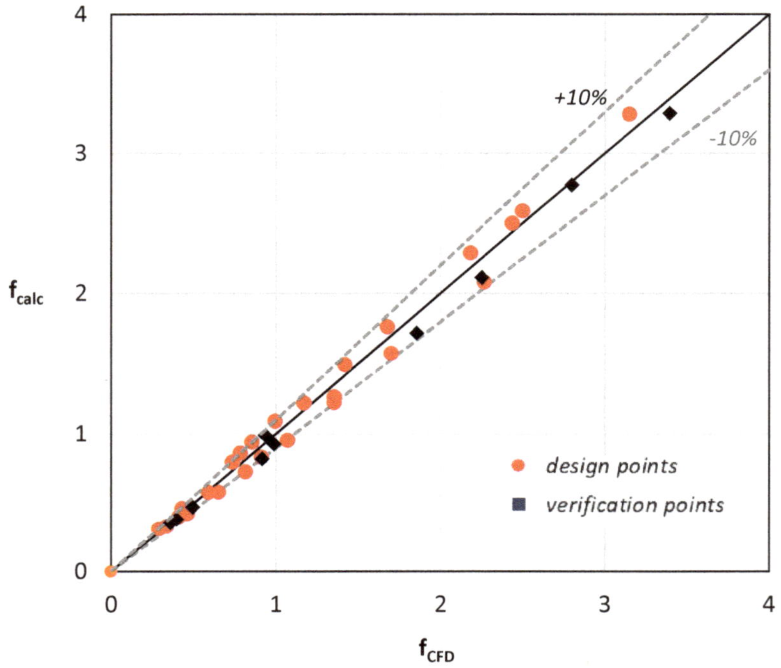

Figure 11. A Comparison between the calculated and predicted values.

5. Concluding Remarks

In this study, a simplified model that can predict Fanning friction factors of laminar blood flow in small-caliber vessels was proposed. The study was conducted using CFD simulations that were validated with relevant experimental data acquired by employing an appropriate experimental setup. The effect of the flow rate, the hematocrit value and the vessel diameter on the pressure drop was considered. The interpretation of the resulting data led to the proposal of a straightforward method of estimating with reasonable accuracy (i.e., better than ±10%) the Fanning friction factor and consequently, the pressure drop during the flow of blood (for a hematocrit range 35–55%) in small-diameter vessels (300–1800 μm).

The calculation procedure comprises the following steps:

1. For a given volumetric flow rate (Q) and vessel inside diameter (D), the pseudo-shear rate (γ^*) is calculated using Equation (5).
2. For a given hematocrit value (H_t), an effective viscosity (μ^*), that corresponds to the pseudo-shear rate is estimated using Equations (1)–(3).
3. The corresponding Re^* and Bm numbers are calculated by Equations (4) and (9).
4. The Fanning friction factor (f) is then calculated by the proposed correlation (Equation (10)).
5. Finally, the pressure drop ($\Delta P/L$) is calculated using Equation (6).

The proposed methodology (Figure 12) is an easy-to-use tool that can help scientist to quickly and accurately estimate the pressure drop exerted during blood flow in healthy, small caliber vessels.

It is clear, that this methodology is not exclusively applicable for blood flow, however, if step two is excluded, it can be applied to any shear thinning Casson fluid.

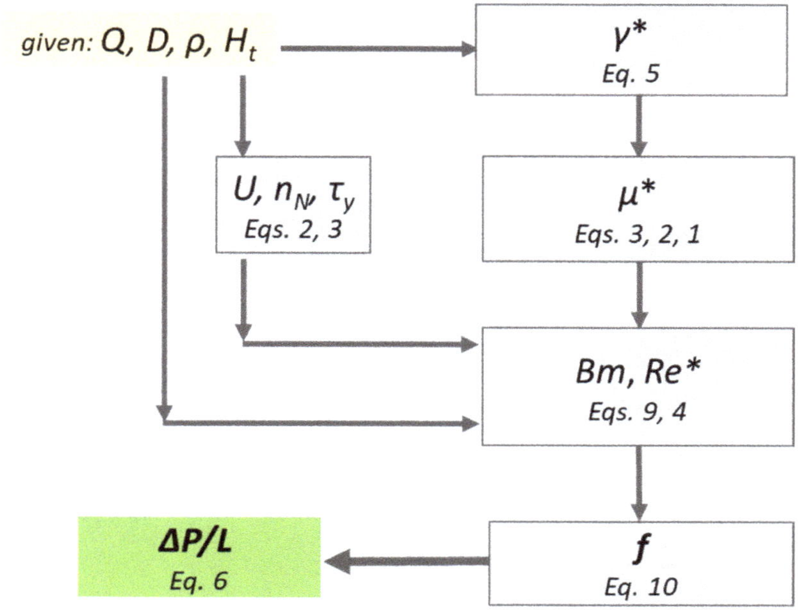

Figure 12. Calculation procedure of the proposed methodology.

Author Contributions: S.V.P. had the initial conception of this work; S.V.P. designed the experiments; I.D.T. performed the experiments; A.A.M. designed the CFD simulations; O.D.S. performed the simulations; A.A.M. and S.V.P. interpreted the numerical results; A.A.M. wrote the manuscript.

Funding: This research received no external funding.

Acknowledgments: The authors would like to thank the laboratory technician, Asterios Lekkas, for the construction and installation of the experimental setup.

Conflicts of Interest: The authors declare no conflict of interest.

Nomenclature

Bm	Bingham number, -
D	Inside vessel diameter, m
f	Fanning friction factor, -
f_{CFD}	Fanning friction factor from CFD simulations, -
f_{calc}	Fanning friction factor from Equation (7), -
H_t	Hematocrit, %
H_{tc}	Critical hematocrit, %
L	Length, m
n_p	Plasma viscosity, Pa·s
P	Pressure, Pa
Q	Volumetric flow rate, mm^3/s
Re_∞	Reynolds number corresponding to μ_∞
Re^*	Effective Reynolds number (Equation (4)), -
U	Mean velocity, m/s
x	Axial coordinate, m

Greek letters

γ^*	Pseudo shear rate, s^{-1}
ΔP	Pressure drop, Pa
μ	Blood viscosity, Pa·s
μ^*	Effective viscosity, Pa·s
μ_∞	Asymptotic viscosity value, Pa·s
ρ	Blood density, kg/m^3
τ	Shear stress, Pa
τ_y	Yield stress, Pa

References

1. Tang, G.H.; Lu, Y.B.; Zhang, S.X.; Wang, F.F.; Tao, W.Q. Experimental investigation of non-Newtonian liquid flow in microchannels. *J. Non-Newtonian Fluid Mech.* **2012**, *173–174*, 21–29. [CrossRef]
2. Thiriet, M. *Biology and Mechanics of Blood Flows Part II: Mechanics and Medical Aspects*; Springer: Paris, France, 2007.
3. Fournier, R. *Basic Transport Phenomena in Biomedical Engineering*, 2nd ed.; CRC Press: New York, NY, USA, 2006.
4. Chilton, R.A.; Stainsby, R. Pressure loss equations for laminar and turbulent non-Newtonian pipe flow. *J. Hydraul. Eng.* **1998**, *124*, 522–528. [CrossRef]
5. Anastasiou, A.D.; Spyrogianni, A.S.; Koskinas, K.C.; Giannoglou, G.D.; Paras, S.V. Experimental investigation of the flow of a blood analogue fluid in a replica of a bifurcated small artery. *Med. Eng. Phys.* **2011**, *34*, 211–218. [CrossRef] [PubMed]
6. Metzner, A.B.; Reed, J.C. Flow of Non-Newtonian Fluids—Correlation of the Laminar, Transition, and Turbulent-flow Regions. *AlChe J.* **1955**, *1*, 434–440. [CrossRef]
7. Chhabra, R.P.; Richardson, J.F. *Non-Newtonian Flow and Applied Rheology: Engineering Application*, 2nd ed.; Elsevier: Amsterdam, The Netherlands, 2008.
8. Merrill, E.W. Rheology of Blood. *Physiol. Rev.* **1969**, *49*, 863–888. [CrossRef] [PubMed]
9. Cruz, D.A.; Coelho, P.M.; Alves, M.A. A simplified method for calculating heat transfer coefficients and friction factors in laminar pipe flow of non-Newtonian fluids. *J. Heat Transf.* **2012**, *134*, 091703. [CrossRef]
10. Neofytou, P. Comparison of blood rheological models for physiological flow simulation. *Biorheology* **2004**, *41*, 693–714. [PubMed]
11. Kanaris, A.G.; Anastasiou, A.D.; Paras, S.V. Modeling the effect of blood viscosity on hemodynamic factors in a small bifurcated artery. *Chem. Eng. Sci.* **2011**, *71*, 202–211. [CrossRef]
12. Nichols, W.W.; O'Rourke, M.F. *McDonald's Blood Flow in Arteries: Theoretical, Experimental and Clinical Principles*, 5th ed.; Oxford University Press: Oxford, UK, 2005.
13. Pries, A.R.; Secomb, T.W. Blood Flow in Microvascular Networks. In *Microcirculation*, 2nd ed.; Tuman, F.R., Durán, N.W., Ley, K., Eds.; Academic Press: San Diego, CA, USA, 2008.
14. Box, G.E.P.; Hunter, J.S.; Hunter, W.G. *Statistics for Experimenters: Design, Innovation and Discovery*, 2nd ed.; John Wiley and Sons, Inc.: Hoboken, NJ, USA, 2005.

fluids

MDPI

Article

Fluid-Structure Interaction in Abdominal Aortic Aneurysms: Effect of Haematocrit

Yorgos G. Stergiou [1], Athanasios G. Kanaris [2], Aikaterini A. Mouza [1] and Spiros V. Paras [1,*]

1 Department of Chemical Engineering, Aristotle University of Thessaloniki, 54124 Thessaloniki, Greece; gstergiou@auth.gr (Y.G.S.); mouza@auth.gr (A.A.M.)

2 Scientific Computing Department, Rutherford Appleton Laboratory, Didcot OX11 0QX, UK; athanasios.kanaris@stfc.ac.uk

* Correspondence: paras@auth.gr; Tel.: +30-2310-996-174

Received: 14 December 2018; Accepted: 9 January 2019; Published: 14 January 2019

Abstract: The Abdominal Aortic Aneurysm (AAA) is a local dilation of the abdominal aorta and it is a cause for serious concern because of the high mortality associated with its rupture. Consequently, the understanding of the phenomena related to the creation and the progression of an AAA is of crucial importance. In this work, the complicated interaction between the blood flow and the AAA wall is numerically examined using a fully coupled Fluid-Structure Interaction (FSI) method. The study investigates the possible link between the dynamic behavior of an AAA and the blood viscosity variations attributed to the haematocrit value, while it also incorporates the pulsatile blood flow, the non-Newtonian behavior of blood and the hyperelasticity of the arterial wall. It was found that blood viscosity has no significant effect on von Mises stress magnitude and distribution, whereas there is a close relation between the haematocrit value and the Wall Shear Stress (*WSS*) magnitude in AAAs. This *WSS* variation can possibly alter the mechanical properties of the arterial wall and increase its growth rate or even its rupture possibility. The relationship between haematocrit and dynamic behavior of an AAA can be helpful in designing a patient specific treatment.

Keywords: Abdominal Aortic Aneurysm; Fluid-Structure Interaction (FSI); Computational Fluid Dynamics (CFD); haematocrit; pulsatile flow; non-Newtonian

1. Introduction

Abdominal Aortic Aneurysm (AAA) is a cardiovascular disorder that is a cause for serious concern worldwide. AAA is a local dilation of the abdominal aorta, mostly found in the infrarenal segment, usually over 3 cm in diameter and are most of the time reported in men aged 65 or older that are smokers [1]. This problem, like many of the problems health professionals are confronted with, calls for an interdisciplinary approach. Engineers can tackle medical problems by employing methods that are fundamental to engineering practice to comprehend and modify biological systems to assist in the diagnosis and therapy of human diseases.

Little is understood about the complete mechanism of AAA creation. Understanding the phenomena related to the creation of an AAA, progression and behavior are of crucial importance since AAA patients' mortality appears significantly high, especially for developed countries [2]. Namely, an AAA's rupture is lethal in up to 90% of cases [3] but rarely presents any symptoms until its occurrence [1,4] making its detection challenging. Determining AAA's risk factors, or any factors that lead to the generation, growth and evolution of AAAs can lead to its more effective prevention or cure. The maximum aneurysm size is often picked as a rupture risk index, although many small AAAs do also rupture [5]. Stresses on the aneurismal wall are believed to be a better rupture risk index and offers better data for surgical evaluation [6,7]. Aneurysm rupture may occur when this wall stress surpasses the aneurismal wall failure strength.

Risk factors for the development or even rupture of AAAs include multiple biochemical processes occurring in parts of the aortic wall, erosion of the endothelium of the arterial wall [8], along with physiological haemodynamic abnormalities that alter the interaction of the blood flow with the arterial wall. It is accepted that cardiovascular diseases in general—and more specifically, AAAs—are affiliated with blood viscosity [9]. The current study investigates the possible link between dynamic behavior of an AAA and changes in blood viscosity due to variations of the haematocrit (H_t), which in turn is defined as the volumetric percentage of red blood cells in blood [10]. This association between blood viscosity and AAA haemodynamics rarely appears in literature [11]. Inserting a clinical parameter in the study enables us to correlate the AAA's behavior with a common medical index. Previous research by Kanaris et al. [12] revealed that variations of blood haematocrit can turn out to be noteworthy in haemodynamics (e.g., *WSS*).

In this work the complicated interaction between the blood flow and the AAA wall is numerically examined by using a fully coupled Fluid-Structure Interaction (FSI) method. It is common place that coupling the fluid dynamics component of the simulation with the solid domain simulation is essential to reach more representative results of the overall AAA behavior [13–16]. In the past, the application of CFD (without FSI) in blood flow simulation in patient-specific geometries has led to significant advances in the understanding of how the haemodynamic quantities of interest affect or are affected by the vessel wall geometry and boundary conditions [17]. Over the last decade CFD was successfully used in the simulation of blood flow, examination of potential surgical treatment options, simulation of medical devices, etc. However, CFD-only blood flow always incorporates the assumption that the blood vessel walls are rigid, which is not always a realistic approach. As vascular walls are flexible, they tend to deform due to haemodynamic forces; wall deformation would then alter the blood flow patterns, which in turn alter the fluid dynamic itself. This behavior makes the study of an AAA a purely FSI problem.

Wall Shear Stress (*WSS*) should also be considered when studying AAAs. *WSS* is a significant parameter of haemodynamics as it can alter the arterial wall properties [6]. These alterations may have great impact on the AAA wall as they can reduce its resistance and consequently accelerate its rupture [18,19]. When simulating blood flows, the importance of imposing a pulsatile boundary condition for the mass flow is proved [20] to play key part in subsequent results, especially in dynamic phenomena like AAA wall behavior.

Previous research has revealed that the non-Newtonian nature of blood shall not be ignored as it plays a great role in determining various characteristics of the flow, predominantly [12,21,22]. The present study incorporates a non-Newtonian model for blood using the Casson model [10], integrating simultaneously with the viscosity modelling the haematocrit dependency on the AAA behavior [10]. Summarizing the above, we conclude that while existing literature does consider some of the aforementioned issues, such as the following,

- the pulsatile blood flow [20],
- the non-Newtonian behavior of blood [21],
- the hyperelasticity of the arterial wall [23,24],

in most of the AAA studies, not all of them are included simultaneously.

This study incorporates all the above issues to investigate the possible link between the dynamic behavior of an abdominal aortic aneurysm (AAA) and the blood viscosity variations due to different haematocrit values. This is accomplished by using simulation tools that implement a Fluid-Structure Interaction (FSI) method. Specifically, for a typical range of haematocrit values we will numerically estimate the values of blood pressure, the equivalent stress (von Mises stress or equivalent tensile stress) of the AAA wall and the *WSS*.

2. Methodology

2.1. About FSI

Multiphysics simulation is a crucial tool, which attempts to accurately predict complex phenomena that will occur where multiple types of coupled physics (structure, fluid, thermal etc.) interact. It is a well-established method in product engineering, as it drives development processes and can influence engineering simulation efforts, as its strategic value is being recognized.

Those multiphysics problems are usually solved with a multi-field approach, where physics are treated each as an independent field with its own variables. Different fields "interact" by exchanging information via "interfaces", which are special types of boundary conditions. One very well-known type of multiphysics problems is the Fluid-Structure Interaction (FSI) which occurs when a fluid interacts with a solid structure causing deformation in the structure and, as a result, altering the flow of the fluid itself. A solution based on FSI is required for biomedical flows involving compliant blood vessels and valves.

A coupled system is defined as one where physically or computationally heterogeneous mechanical components interact dynamically [17]. This system is a group of functionally related components which are forming a "collective entity". FSI is focused on understanding the system response due to an excitation or change in boundary conditions. Different physical phenomena associated with the system components usually are nonlinear and act on different time scales and spatial domains. Consequently, each component needs to be described by appropriate theoretical models along with the inherent coupling mechanisms between different fields.

To effectively approach these types of complex systems, it is required that they are "broken-down" or partitioned. Partitioning is the process of spatial separation, "decomposition" of a discrete model into interacting components referred to as "partitions". Decomposition is usually driven by physical or functional considerations: a physical subsystem can be the fluid—the blood—which can be approached by field equations in computational fluid dynamics (CFD) and another physical subsystem can be the blood vessel, which will be approached as a structural model. In a coupled multiphysics problem, multiple physics models or phenomena are handled simultaneously. In this case, different discretization techniques are applied on individual subdomains on different spatial domains, or individual field variables (e.g., pressure applied by blood on a vessel wall) represent different but mutually interacting physical phenomena.

2.2. Geometry

The AAA studied is positioned just below the renal bifurcation. Normal aortic diameters vary widely and are strongly influenced by several factors (e.g., gender, age etc.) [25,26]. A simplified geometry was selected as a benchmark to emulate a real-life AAA for this current parametric and methodological study. The non-affected part of the model was assumed to have a diameter of 20 mm [25–27]. The aortic sac was modelled asymmetrically to match representative abnormalities. The maximum aneurismal diameter was set to 55 mm, which falls in the lower threshold zone for high rupture risk and subsequently is a strong candidate for surgical treatment [28,29]. The total length of the geometry designed is $L = 90$ mm. ANSYS DesignModeler (ANSYS, Inc., Canonsburg, PA, USA) was used to create the simplified representation of an AAA. The AAA model geometry used in the simulations is presented in Figures 1 and 2, while the geometric dimensions of the simplified AAA model are shown in Table 1.

Table 1. Geometrical parameters of the Abdominal Aortic Aneurysm (AAA) model.

Parameter	Value
Total length of the AAA, L	90 mm
Internal inlet diameter of the AAA, D_i	20 mm
Maximum aneurismal diameter, D_{max}	55 mm
Arterial thickness, k	2 mm

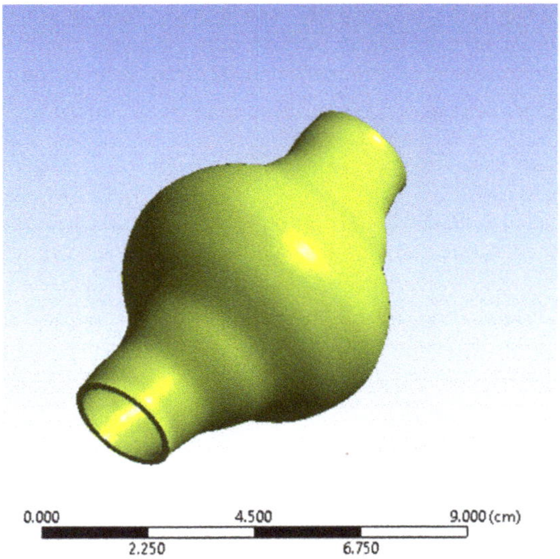

Figure 1. View of the Abdominal Aortic Aneurysm (AAA) wall (solid domain).

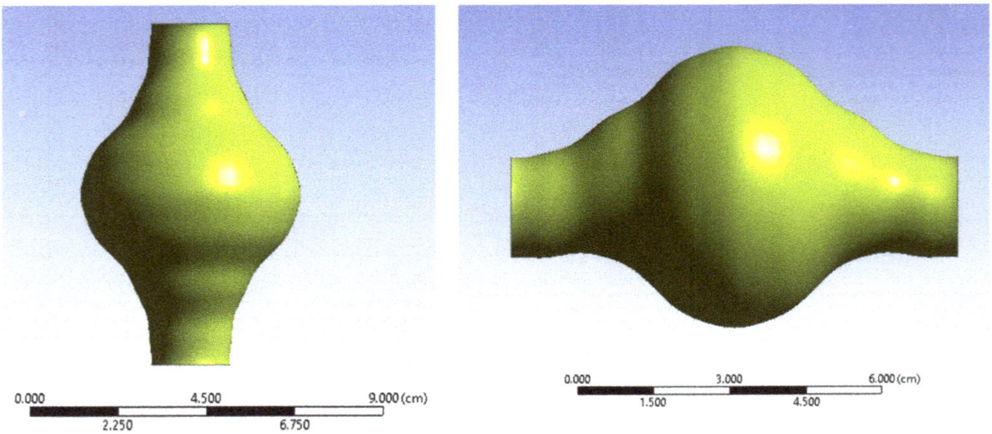

Figure 2. Two views of the AAA fluid domain.

2.3. Governing Equations and Boundary Conditions

The governing equations for the fluid flow are the Navier-Stokes equations for incompressible non-Newtonian flow. The momentum equations (Equation (1)) are expressed in the Arbitrary Lagrangian-Eulerian form (ALE) for the fluid domain [23,30]:

$$\rho_b \frac{\partial u}{\partial t} + \rho_b \left((u - \dot{d}_b) \cdot \nabla \right) u - \nabla \cdot \tau_b = f_b \tag{1}$$

where $\dot{d}_b$ is the velocity vector of the moving mesh boundary interface, ρ_b is the blood density, 1050 kg/m^3, and f_b the body forces per unit volume. The fluid stress tensor (τ_b) is defined as Equation (2):

$$\tau_b = -P\boldsymbol{I} + 2\mu \boldsymbol{D}(\boldsymbol{u}) \tag{2}$$

where $\boldsymbol{D}(\boldsymbol{u})$ is the strain rate tensor and μ the dynamic viscosity. The strain rate tensor is expressed via Equation (3):

$$\boldsymbol{D}(\boldsymbol{u}) = \frac{1}{2}\left(\nabla \boldsymbol{u} + \nabla \boldsymbol{u}^T \right) \tag{3}$$

and $\dot{\gamma}$ is the shear rate defined as Equation (4):

$$\dot{\gamma} = \sqrt{\frac{1}{2}\boldsymbol{D}(\boldsymbol{u}) : \boldsymbol{D}(\boldsymbol{u})} \tag{4}$$

So, Equation (2) can be rewritten as follows:

$$\tau_b = -P\boldsymbol{I} + 2\mu(\dot{\gamma})\boldsymbol{D}(\boldsymbol{u}) \tag{5}$$

Among the various proposed models for relating blood viscosity and shear rate [10], the Casson model was preferred since its constants can be expressed as a function of haematocrit, H_t [31]. The viscosity of blood, $\mu(\dot{\gamma})$, is modelled via Equation (6) [32]:

$$\mu = \left(\sqrt{\frac{\tau_y}{\gamma}} + \sqrt{\mu_\infty} \right)^2 \tag{6}$$

where τ_y is the yield stress and μ_∞ is the asymptotic viscosity value, characteristic for high shear rate values. The yield stress is a measure of the relative resistance of blood to the flow at very low shear rate values, caused by red blood cell aggregates [10]. It is known [10] that there is a strong correlation between blood viscosity and haematocrit (H_t). A proposed model, which relates the aforementioned terms of Equation (6) with the value of H_t [10], is used in this case:

$$\mu_\infty = \mu_p \left[1 + 0.025 H_t + 7.35 \cdot 10^{-4} \cdot H_t^2 \right] \tag{7}$$

where μ_p is the viscosity of blood plasma, and

$$\tau_y = A(H_t - H_{tc})^3 \tag{8}$$

H_{tc} is the critical haematocrit value, below which the yield stress (τ_y) influence becomes insignificant. Usually, H_{tc} ranges between 4 and 8, whereas A is a constant that ranges between 0.6×10^{-7} and 1.2×10^{-7} Pa. In this study, the values selected for A and H_{tc} were 0.9×10^{-7} Pa and 6, respectively (Figure 3).

As for the solid domain, i.e., the arterial wall, the governing equation follows the movement of the solid material on a moving coordinate system. The solid elastodynamics are described by Equation (9):

$$\nabla \cdot \tau_W + f_W = \rho_W \ddot{d_W} \quad (9)$$

where τ_W is the stress tensor on the arterial wall, f_W, the arterial wall force per unit volume, ρ_W, the arterial wall density, 2000 kg/m^3 and $\ddot{d_W}$, the arterial wall local acceleration. The model used to describe the arterial wall properties was the Mooney-Rivlin model [33,34] that perceives the arterial wall as a nonlinear, isotropic and hyperelastic material using a simplified model of the strain density function (Equation (10)):

$$\Psi = C_1(_1 - 3) + C_2(I_1 - 3)^2 \quad (10)$$

where Ψ is the strain energy, I_1 the first invariant of the left Cauchy-Green tensor and the values C_1 = 17.4 N/cm^2, C_2 = 188.1 N/cm^2 were obtained from Raghavan & Vorp [35].

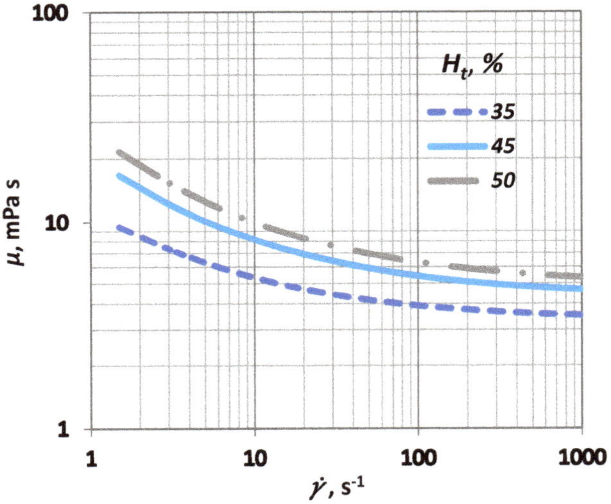

Figure 3. Influence of shear rate ($\dot{\gamma}$) on blood viscosity (μ) for various H_t values.

For boundary conditions, a pulsatile inlet was reproduced from [36] (Figure 4). Additionally, a corresponding pulsatile pressure profile was implemented at the outlet as a normal traction boundary condition (Figure 5) [36]. The heart rate period was T = 1 s and the Womersley number, calculated for the inlet, was $Wo = 14.9$, a normal value for a human aorta in normal conditions [37]. The solid domain was fixed at the inlet and the outlet and it was assumed that this immobility does not influence the AAA wall displacement significantly. On the outer surface of the arterial wall the absolute pressure was set to atmospheric.

As for the FSI boundary conditions, it was assumed that the displacement of the interface is the same for the fluid and solid domains. For the fluid part, the interface was designated as a no-slip wall. The interaction on the surface is described by Equations (11)–(13) [38]:

$$d_W = d_b \quad (11)$$

$$n \cdot \sigma_W = n \cdot \sigma_b \quad (12)$$

$$\dot{d_W} = \dot{d_b} \tag{13}$$

where d is the displacement for each domain and n is the normal vector of the corresponding boundary surface, along with the corresponding subscripts w, for the arterial wall and b for the blood flow.

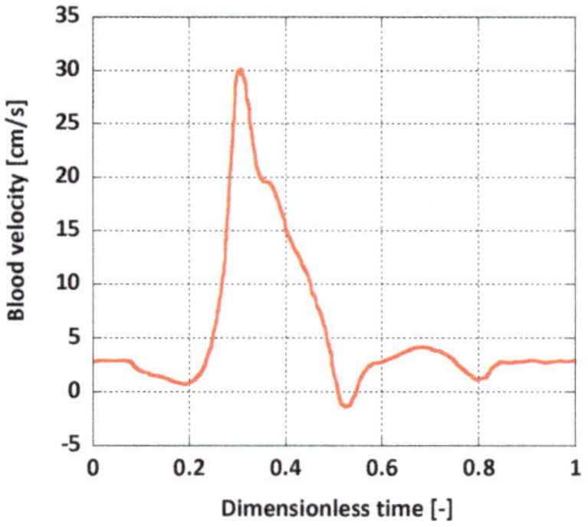

Figure 4. Inlet boundary condition.

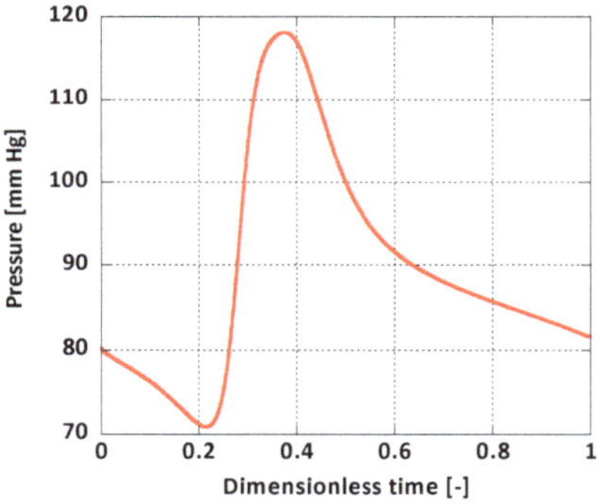

Figure 5. Outlet boundary condition.

3. CFD Modelling

The coupled FSI simulations were performed using a commercial CFD code, the ANSYS Workbench® software (v. 19, ANSYS Inc., Canonsburg, PA, USA). More precisely, the fluid domain was solved using ANSYS CFX® and the solid domain was solved in ANSYS Mechanical. The coupling was performed by the ANSYS Workbench® coupling component. At the beginning of each time step, ANSYS CFX calculates the field variables for the fluid domain and passes the resulting pressure loads on the interface to the ANSYS Mechanical, which in turn solves the finite element model (FEM) for the solid domain. These stagger loops in the same timestep are reinitialized with the new deformed mesh that occurs from the solid deformation (Figure 6).

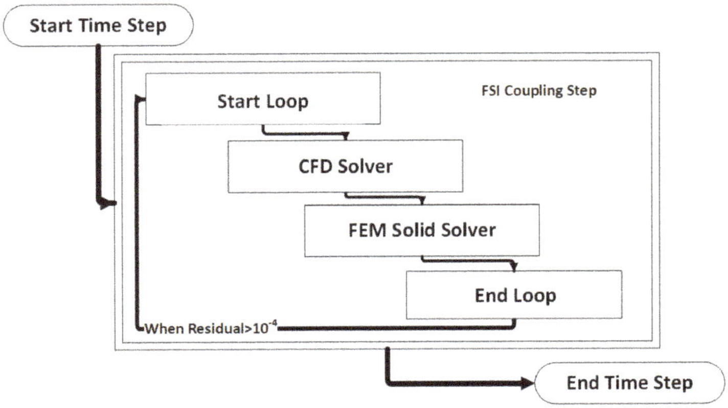

Figure 6. Visual representation of an FSI coupling timestep.

Hexahedral elements were used for the discretization of the fluid domain shown in Figure 7. Adequate inflation was applied near the wall. The solid domain was discretized using 20-node elements suitable for hyperelastic modelling in a one-layer layout, (Figure 8) as this is common practice in similar solid mechanical models. Optimum grid density for the fluid domain was selected by performing a grid dependency study. The fluid domain was discretized in 402,800 cells.

For the fluid domain, the finite volume method and a fully coupled solver for the pressure and velocity, provided by ANSYS CFX®, are used. The number of iterations ensure that mass and momentum residual values are less than 10^{-12}, while the data transfer between the two FSI components continues until the relevant residuals reached an acceptable value (i.e., 10^{-4}). The *DNS* method for laminar flow was employed for the solution, as the flow in the AAA demonstrated no turbulent characteristics (Re_{max} = 1900) [39]. For the space and time integration, the second-order upwind and backward Euler methods were used, respectively, whereas for the solid domain a three-dimensional Finite Element Method (FEM) was employed. The physical time step (Δt) was set to a constant value, which in our case was Δt = 0.0005 s to comply with the selected convergence criteria. All simulations were performed in a custom-made cluster for parallel computing (HPC) consisting of 32 AMD Opteron (AMD Inc., Santa Clara, CA, USA) cores and 128 GB RAM. The simulation time required for one full cardiac pulse was about 50 h.

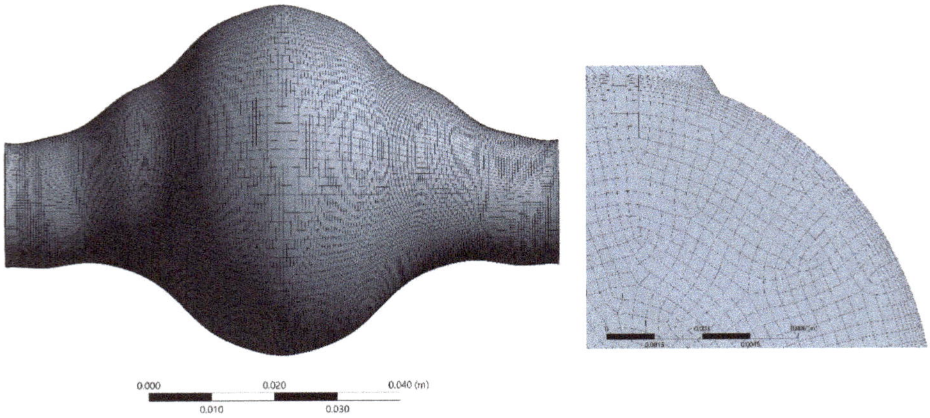

Figure 7. Space discretization of the fluid domain.

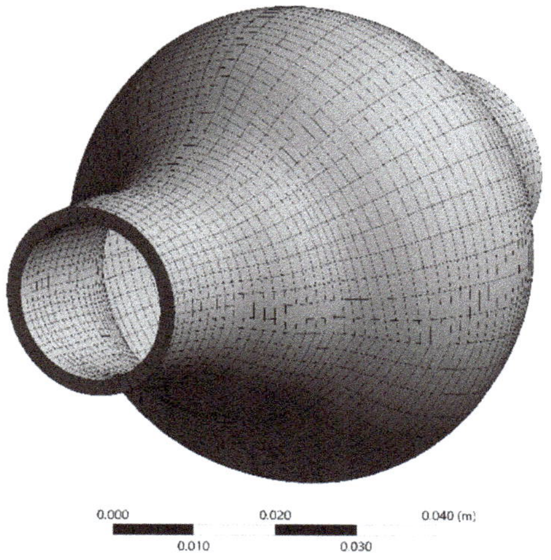

Figure 8. Space discretization of the solid domain.

Several H_t values were tested. In this study, results of two extreme haematocrit values (30%, 50%) are presented and discussed.

4. Results

4.1. Blood flow in the AAA

The *FSI* simulations provide both qualitative and quantitative results that predict blood flow patterns in the AAA (Figure 9). Blood flow in the AAA sac gets significantly decelerated due to the enlarged diameter of the vessel. The temporal variation of blood velocity in the aneurysm follows the inlet boundary velocity profile. The maximum velocity value on the plane cutting the AAA in half at the level of the largest diameter is one order of magnitude lower than the maximum velocity at the inlet. Reverse flow and recirculation zones are present during a full pulse. The pressure distribution in the AAA is highly uniform and oscillates between the two extreme boundary values (Figure 6). The pressure drop between the two endings of the AAA is minimal (~30 Pa) for both cases.

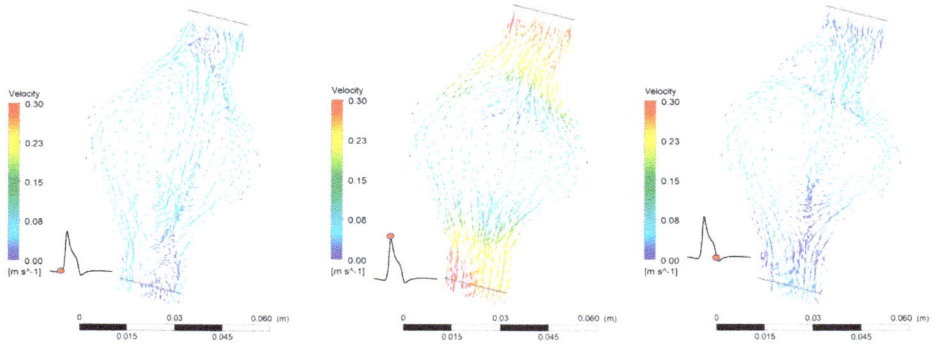

Figure 9. Flow patterns in the AAA during one pulse for H_t = 30%.

4.2. AAA Wall Displacement

Results revealed that total displacement of the aneurismal wall is strongly depended on location. Namely, as shown in Figure 10, displacement is bigger near the areas where diameter variations are more pronounced. Maximum displacement during one pulse occurs when $t/T = 0.373$. No differences appear for the total wall displacement when the results between the two H_t cases are compared. The maximum displacement is 3.90 mm for both cases.

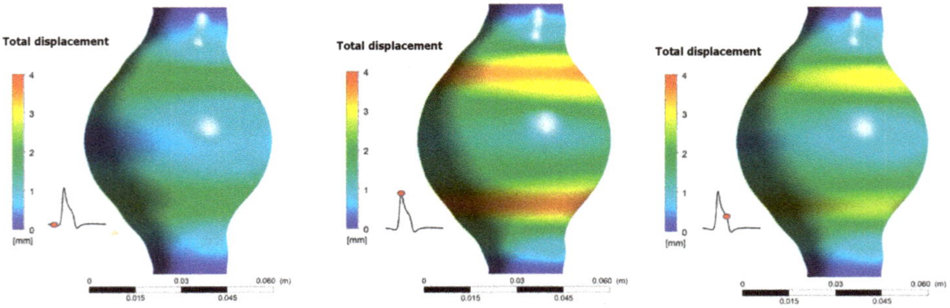

Figure 10. Total displacement of the aneurismal wall during one pulse for H_t = 30%.

4.3. von Mises Stress (σ_{vM}) Distribution on AAA Wall

As expected, there is a strong correlation between the arterial wall displacement and the distribution of von Mises stress. Von Mises stress is a value used to predict yielding of materials under complex loading and it can be represented in an expression that uses the different components of a stress tensor. As stated in the literature [40], the von Mises stress, σ_{vM}, can be calculated by Equation (14):

$$\sigma_{vM} = \sqrt{\frac{1}{2}[(\sigma_1 - \sigma_2)^2 + (\sigma_2 - \sigma_3)^2 + (\sigma_3 - \sigma_1)^2]} \tag{14}$$

where σ_1, σ_2 and σ_3 are the principal stresses in three-dimensional problems.

It is revealed (Figure 11) that peak σ_{vM} values appear at the area of maximum deformation of the arterial. Maximum values for both cases occur at $t/T = 0.373$. Peak σ_{vM} appears in the same regions as maximum displacement appears. It was again revealed that, for the two H_t values tested, the σ_{vM} magnitude was not affected. For both H_t = 30% and H_t = 50% the peak σ_{vM} value was 219.4 kPa. Moreover, for the areas that experience the highest stresses, σ_{vM} nearly doubles in the course of one pulse cycle.

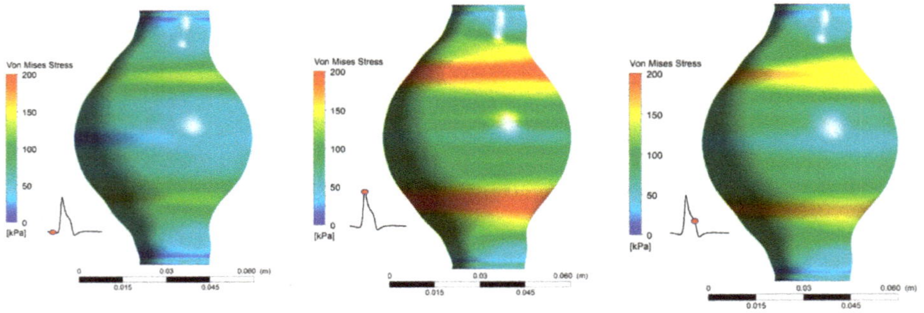

Figure 11. von Mises Stress distribution during one pulse for H_t = 30%.

By taking the average value of σ_{vM} on the arterial wall during one full pulse cycle, it is possible to visually compare the differences on wall dynamics caused by the change in blood viscosity due to H_t variations (Figure 12). As deduced from Figure 12, the average σ_{vM} is practically the same for the two H_t cases tested.

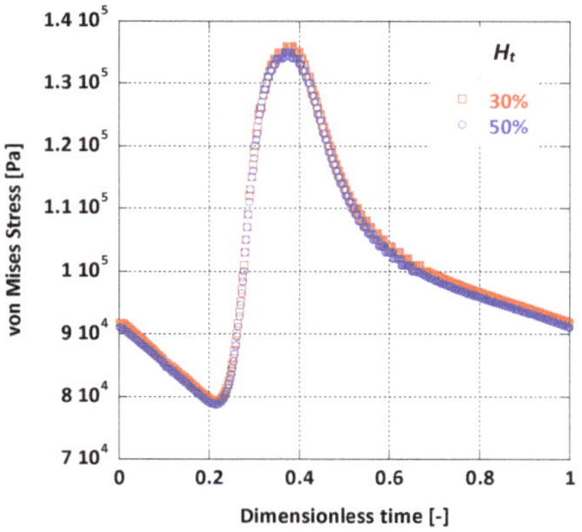

Figure 12. Effect of H_t on the surface averaged σ_{vM} on the AAA wall.

4.4. *WSS in the AAA*

WSS values vary significantly between the two cases tested and are highly influenced by both the location in the AAA and the elapsed time. It is also found that *WSS* depends strongly on blood viscosity, as an increase in H_t causes an increase in calculated peak *WSS* value. The results show that H_t greatly affects *WSS* magnitude (Figure 13). It is notable that near the AAA endings, where the aortic diameter has normal values, *WSS* attains values over 0.7 Pa, which may, according to [12], be in the normal range of *WSS* for healthy individuals. It is evident that low *WSS* values are caused by the relatively slow flow of the blood in the AAA. This low flow velocity appears in areas where the diameter gets larger than normal.

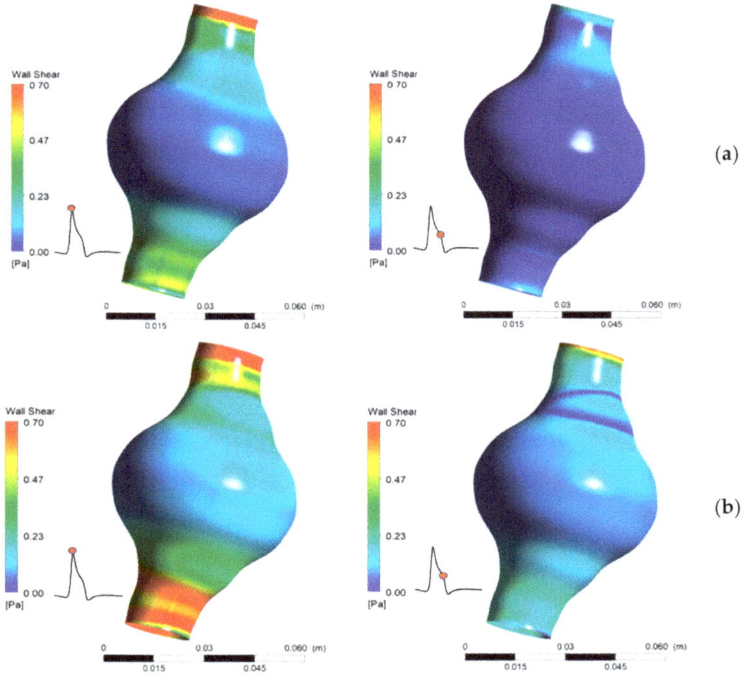

Figure 13. Comparison of *WSS* for two values of H_t (**a**) 30%, (**b**) 50%.

When comparing the area averaged *WSS* values for the two H_t cases (Figure 14) it is notable that there is a major difference in *WSS* magnitude. For the lower H_t value the average *WSS* does not climb over 0.3 Pa during the whole pulse cycle, whereas for H_t = 50%, *WSS* always appears remarkably higher. It is obvious that a nearly 40% decrease in haematocrit can probably cause as much as 70% decrease in *WSS* values. This is also evident by another risk index, the mean time-averaged *WSS* (*TAWSS*); for the AAA in this model and for H_t = 50%, *TAWSS* = 0.16 Pa, while for H_t = 30%, *TAWSS* drops to 0.06 Pa.

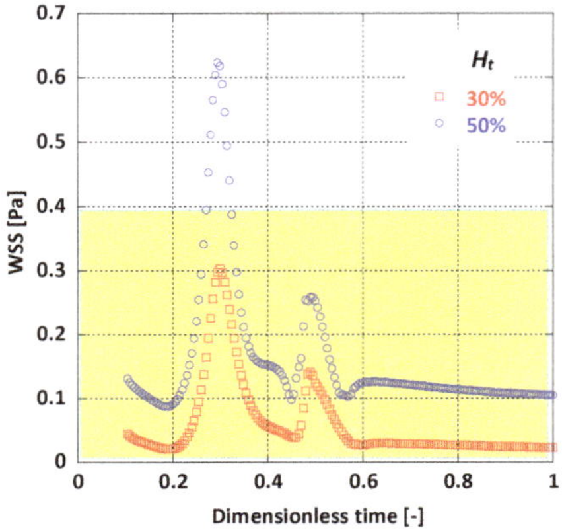

Figure 14. Effect of H_t on the average *WSS*.

5. Discussion

This study proposes a possible link between haematocrit variations in blood and *WSS* values in AAAs.

It is reported [12,18] that *WSS* values less than 0.4 Pa (Figure 15) can result in plaque build-up in the arterial wall. This, in turn, can cause serious abnormalities in the arterial wall's physical and mechanical properties, resulting in an increase of AAA's growth rate, rise in its rupture risk, or change of its post-surgery behavior. The negative outcome of low *WSS* values on the arterial wall is a result of complex mechanical and biochemical phenomena, possibly linked to atherosclerosis [41–44]. More precisely, O'Leary et al. [45] imply causal relation between plaque affected areas and high rupture risk in AAAs. This interaction might be responsible for the adverse clinical situation of patients suffering from AAAs and, at the same time, have lower than normal H_t [9].

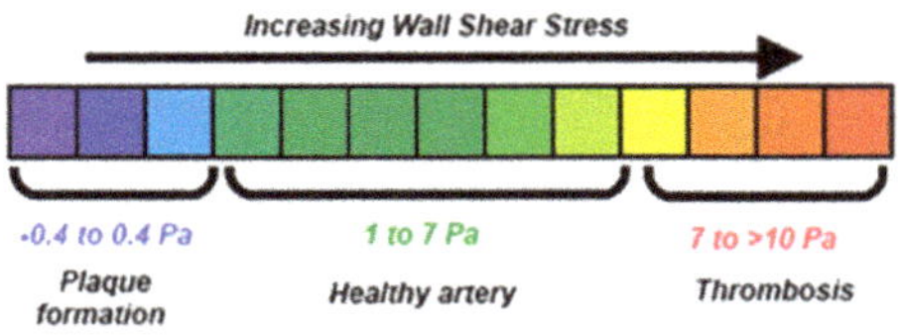

Figure 15. Variations of *WSS* in human arteries [18].

This hypothesis is widely accepted in literature; namely, the mechanism of generation and progression of AAAs, and even its rupture risk can be linked to low-*WSS* induced alterations of the arterial wall [46–50].

Blood viscosity has a clear effect on *WSS* as depicted in the results section. Wang & Li [11] have examined the influence of blood viscosity in AAAs and their conclusions agree qualitatively with the present study.

Overall, this study presents results that are in agreement with previous work on AAAs. Namely, von Mises stress values, along with *WSS* in the aneurismal sac, coincide qualitatively with previous computational research [14,51]. Furthermore, a qualitative agreement exists with clinical research on AAAs regarding the total displacement of the aneurismal wall during one pulse [52].

The present study, like any computational study, includes simplifications and is bound by their limitations. Firstly, an idealized AAA shape with a single and uniform wall thickness was chosen for the sake of simplicity instead of an actual, highly asymmetric patient-specific AAA geometry. The reasoning behind this decision was to avoid introducing highly non-linear effects caused by the imperfections of a real artery, which would in turn introduce noise in the results and make drawing conclusions harder. Having said that, further research is needed to reveal the effect of patient-specific geometric irregularities (combined with the haematocrit changes) on the results of this study (i.e., *WSS* distribution and magnitude, von Mises stresses). As for the arterial wall modelling, a uniform, isotropic, single-layer material was used; this assumption inserts an extra limitation to the model. While this is not realistic, the introduction of more complex materials would not change the effect of H_t, only its magnitude. The effect of this limitation can get minimized by including an even more realistic arterial wall model (e.g., multi-layered approach for the arterial wall) and possibly run a Design of Experiments (DOE) set of simulations to better understand the interaction, which however is out of the scope of this work. For similar reasons, this study considers that, regardless of the values of haematocrit in blood, the pulsatile flow of remains identical, in magnitude and frequency. Moreover, the AAA was positioned at a significant distance downstream from the aortic arch so that any secondary swirling flow is considered negligible. Lastly, fixed artery inlet and outlet were considered in this model for computational simplicity; alternatively, a different boundary condition on the solid domain could be used.

6. Conclusions

This work provides a qualitative insight on the way haematocrit could affect AAA's mechanics and haemodynamics. It takes into account the pulsatile blood flow, the non-Newtonian behavior of the blood and the hyperelasticity of the arterial wall. Coupled CFD & FEM simulations revealed that the variation in blood viscosity does not have a significant effect on AAA's wall solid dynamics as well as on σ_{vM} magnitude and distribution. However, our results show that there is a strong relation between H_t value and the shear stress acting on the arterial wall. As illustrated in Figure 16, lower H_t values result in lower viscosity values and consequently in lower *WSS* values, which in turn promote plaque formation on the aneurismal wall. This could be one of the causal paths describing the effect of low haematocrit values on AAA morbidity.

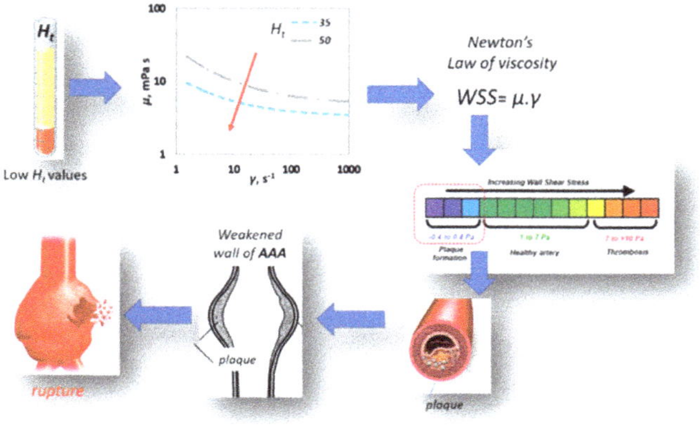

Figure 16. Effect of H_t on AAA rupture.

Further research is needed in the investigation of the link between H_t and the behaviour of AAAs. This FSI approach could be applied in patient specific treatment in order to offer a more robust assessment on the phenomena that may relate low H_t with adverse AAA morbidity.

Author Contributions: S.V.P. had the initial conception of this work and organized it; Y.G.S. designed the CFD/FSI simulations acquired and analyzed the data and interpreted the results; A.G.K. has provided insights on the improvement of computational performance for the required simulations; Y.G.S., A.A.M. and S.V.P. drafted the paper; A.A.M. and S.V.P. reviewed and edited the manuscript.

Funding: This research received no external funding.

Conflicts of Interest: All authors state that there is no conflict of interest. This work is not affiliated with STFC.

Nomenclature

A	Yield stress constant, Pa
C_1	Mooney-Rivlin constant 1, N/cm^2
C_2	Mooney-Rivlin constant 2, N/cm^2
D_{max}	Maximum aneurismal diameter, mm
D_i	Internal inlet AAA diameter, mm
d	*FSI* interface displacement, m
$\dot{d}$	*FSI* interface velocity, m/s
$\ddot{d}$	*FSI* interface acceleration, m/s^2
f	Arterial wall force per volume, N/m^3
H_t	Haematocrit, %
k	Arterial wall thickness, mm
L	Total length of the AAA, mm
P	Pressure, Pa
Re	Reynolds number, dimensionless
t	Time, s
T	Heart rate period, s
u	Velocity, m/s
W_o	Womersley number, dimensionless
WSS	Wall shear stress, Pa
Greek letters	
$\dot{\gamma}$	Shear rate, s^{-1}
Δt	Timestep, s
μ	Viscosity, Pa·s
ρ	Density, kg/m^3
σ_{vM}	von Mises stress, Pa
τ	Shear stress, Pa
τ_y	Yield stress, Pa
Ψ	Strain energy density, J/m^3

References

1. Keisler, B.; Carter, C. Abdominal aortic aneurysm. *Am. Fam. Physician* **2015**, *91*, 538–543. [PubMed]
2. Singh, K.; Bønaa, K.H.; Jacobsen, B.K.; Bjørk, L.; Solberg, S. Prevalence of and risk factors for abdominal aortic aneurysms in a population-based study: The Tromsø Study. *Am. J. Epidemiol.* **2001**, *154*, 236–244. [CrossRef] [PubMed]
3. Assar, A.N.; Zarins, C.K. Ruptured abdominal aortic aneurysm: A surgical emergency with many clinical presentations. *Postgrad. Med. J.* **2009**, *85*, 268–273. [CrossRef] [PubMed]
4. Chervu, A.; Clagett, G.P.; Valentine, R.J.; Myers, S.I.; Rossi, P.J. Role of physical examination in detection of abdominal aortic aneurysms. *Surgery* **1995**, *117*, 454–457. [CrossRef]
5. Vorp, D.A. Biomechanics of abdominal aortic aneurysm. *J. Biomech.* **2007**, *40*, 1887–1902. [CrossRef] [PubMed]

6. Fillinger, M.F.; Marra, S.P.; Raghavan, M.L.; Kennedy, F.E. Prediction of rupture risk in abdominal aortic aneurysm during observation: Wall stress versus diameter. *J. Vasc. Surg.* **2003**, *37*, 724–732. [CrossRef] [PubMed]
7. Georgakarakos, E.; Ioannou, C.V.; Kamarianakis, Y.; Papaharilaou, Y.; Kostas, T.; Manousaki, E.; Katsamouris, A.N. The Role of Geometric Parameters in the Prediction of Abdominal Aortic Aneurysm Wall Stress. *Eur. J. Vasc. Endovasc. Surg.* **2010**, *39*, 42–48. [CrossRef] [PubMed]
8. Xenos, M.; Rambhia, S.H.; Alemu, Y.; Einav, S.; Labropoulos, N.; Tassiopoulos, A.; Ricotta, J.J.; Bluestein, D. Patient-Based Abdominal Aortic Aneurysm Rupture Risk Prediction with Fluid Structure Interaction Modeling. *Ann. Biomed. Eng.* **2010**, *38*, 3323–3337. [CrossRef]
9. Diehm, N.; Benenati, J.F.; Becker, G.J.; Quesada, R.; Tsoukas, A.I.; Katzen, B.T.; Kovacs, M. Anemia is associated with abdominal aortic aneurysm (AAA) size and decreased long-term survival after endovascular AAA repair. *J. Vasc. Surg.* **2007**, *46*, 676–681. [CrossRef] [PubMed]
10. Errill, E.W. Rheology of blood. *Physiol. Rev.* **1969**, *49*, 863–888. [CrossRef] [PubMed]
11. Wang, X.; Li, X. Computational simulation of aortic aneurysm using FSI method: Influence of blood viscosity on aneurismal dynamic behaviors. *Comput. Biol. Med.* **2011**, *41*, 812–821. [CrossRef] [PubMed]
12. Kanaris, A.G.; Anastasiou, A.D.; Paras, S.V. Modeling the effect of blood viscosity on hemodynamic factors in a small bifurcated artery. *Chem. Eng. Sci.* **2012**, *71*, 202–211. [CrossRef]
13. Scotti, C.M.; Shkolnik, A.D.; Muluk, S.C.; Finol, E.A. Fluid-structure interaction in abdominal aortic aneurysms: Effects of asymmetry and wall thickness. *Biomed. Eng. Online* **2005**, *4*, 64. [CrossRef] [PubMed]
14. Scotti, C.M.; Jimenez, J.; Muluk, S.C.; Finol, E.A. Wall stress and flow dynamics in abdominal aortic aneurysms: Finite element analysis vs. fluid–structure interaction. *Comput. Methods Biomech. Biomed. Eng.* **2008**, *11*, 301–322. [CrossRef] [PubMed]
15. Leung, J.H.; Wright, A.R.; Cheshire, N.; Crane, J.; Thom, S.A.; Hughes, A.D.; Xu, Y. Fluid structure interaction of patient specific abdominal aortic aneurysms: A comparison with solid stress models. *Biomed. Eng. Online* **2006**, *5*, 33. [CrossRef]
16. Lin, S.; Han, X.; Bi, Y.; Ju, S.; Gu, L. Fluid-Structure Interaction in Abdominal Aortic Aneurysm: Effect of Modeling Techniques. *BioMed. Res. Int.* **2017**, *2017*, 7023078. [CrossRef]
17. Bazilevs, Y.; Takizawa, K.; Tezduyar, T.E. *Computational Fluid-Structure Interaction: Methods and Applications*; John Wiley & Sons: Hoboken, NJ, USA, 2013; ISBN 978-1-118-48357-2.
18. Malek, A.M.; Alper, S.L.; Izumo, S. Hemodynamic shear stress and its role in atherosclerosis. *JAMA* **1999**, *282*, 2035–2042. [CrossRef]
19. Boyd, A.J.; Kuhn, D.C.S.; Lozowy, R.J.; Kulbisky, G.P. Low wall shear stress predominates at sites of abdominal aortic aneurysm rupture. *J. Vasc. Surg.* **2016**, *63*, 1613–1619. [CrossRef]
20. Finol, E.A.; Amon, C.H. Blood Flow in Abdominal Aortic Aneurysms: Pulsatile Flow Hemodynamics. *J. Biomech. Eng.* **2001**, *123*, 474–484. [CrossRef]
21. Khanafer, K.M.; Gadhoke, P.; Berguer, R.; Bull, J.L. Modeling pulsatile flow in aortic aneurysms: Effect of non-Newtonian properties of blood. *Biorheology* **2006**, *43*, 661–679.
22. Johnston, B.M.; Johnston, P.R.; Corney, S.; Kilpatrick, D. Non-Newtonian blood flow in human right coronary arteries: Transient simulations. *J. Biomech.* **2006**, *39*, 1116–1128. [CrossRef] [PubMed]
23. Mesri, Y.; Niazmand, H.; Deyranlou, A.; Sadeghi, M.R. Fluid-structure interaction in abdominal aortic aneurysms: Structural and geometrical considerations. *Int. J. Mod. Phys. C* **2014**, *26*, 1550038. [CrossRef]
24. Sharzehee, M.; Khalafvand, S.S.; Han, H.-C. Fluid-structure interaction modeling of aneurysmal arteries under steady-state and pulsatile blood flow: A stability analysis. *Comput. Methods Biomech. Biomed. Eng.* **2018**, *21*, 219–231. [CrossRef] [PubMed]
25. Joh, J.H.; Ahn, H.-J.; Park, H.-C. Reference diameters of the abdominal aorta and iliac arteries in the Korean population. *Yonsei Med. J.* **2013**, *54*, 48–54. [CrossRef] [PubMed]
26. Syed, M.N.; Ahmad, M.M.; Ahmad, M.N.; Hussaini, S.; Muhammad, M.N.; Pir, S.H.A.; Khandheria, B.K.; Tajik, A.J.; Ammar, K.A. Normal Diameter of the Ascending Aorta in Adults: The Impact of Stricter Criteria on Selection of Subjects Free of Disease. *J. Am. Coll. Cardiol.* **2017**, *69*, 2075. [CrossRef]
27. Leotta, D.F.; Paun, M.; Beach, K.W.; Kohler, T.R.; Zierler, R.E.; Strandness, D.E. Measurement of abdominal aortic aneurysms with three-dimensional ultrasound imaging: Preliminary report. *J. Vasc. Surg.* **2001**, *33*, 700–707. [CrossRef] [PubMed]

28. Brown, P.M.; Zelt, D.T.; Sobolev, B. The risk of rupture in untreated aneurysms: The impact of size, gender, and expansion rate. *J. Vasc. Surg.* **2003**, *37*, 280–284. [CrossRef] [PubMed]
29. Lederle, F.A.; Johnson, G.R.; Wilson, S.E.; Ballard, D.J.; William, D.; Jordan, J.; Blebea, J.; Littooy, F.N.; Freischlag, J.A.; Bandyk, D.; et al. Rupture Rate of Large Abdominal Aortic Aneurysms in Patients Refusing or Unfit for Elective Repair. *JAMA* **2002**, *287*, 2968–2972. [CrossRef]
30. Janela, J.; Moura, A.; Sequeira, A. A 3D non-Newtonian fluid–structure interaction model for blood flow in arteries. *J. Comput. Appl. Math.* **2010**, *234*, 2783–2791. [CrossRef]
31. Neofytou, P. Comparison of blood rheological models for physiological flow simulation. *Biorheology* **2004**, *41*, 693–714.
32. Fournier, R.L. *Basic Transport Phenomena in Biomedical Engineering*, 3rd ed.; CRC Press: New York, NY, USA, 2011; ISBN 978-1-4398-2670-6.
33. Mooney, M. A Theory of Large Elastic Deformation. *J. Appl. Phys.* **1940**, *11*, 582–592. [CrossRef]
34. Rivlin, R.S. Large elastic deformations of isotropic materials IV. further developments of the general theory. *Philos. Trans. R. Soc. Lond. A* **1948**, *241*, 379–397. [CrossRef]
35. Raghavan, M.L.; Vorp, D.A. Toward a biomechanical tool to evaluate rupture potential of abdominal aortic aneurysm: Identification of a finite strain constitutive model and evaluation of its applicability. *J. Biomech.* **2000**, *33*, 475–482. [CrossRef]
36. Mills, C.J.; Gabe, I.T.; Gault, J.H.; Mason, D.T.; Ross, J.; Braunwald, E.; Shillingford, J.P. Pressure-flow relationships and vascular impedance in man. *Cardiovasc. Res.* **1970**, *4*, 405–417. [CrossRef] [PubMed]
37. Nichols, W.W.; McDonald, D.A.; O'Rourke, M.F. *McDonald's Blood Flow in Arteries: Theoretical, Experimental and Clinical Principles*, 5th ed.; Taylor & Francis: Milton Park, UK, 2005; ISBN 978-0-340-80941-9.
38. Maday, Y. Analysis of coupled models for fluid-structure interaction of internal flows. In *Cardiovascular Mathematics: Modeling and Simulation of the Circulatory System*; Formaggia, L., Quarteroni, A., Veneziani, A., Eds.; Springer-Verlag: Mailand, Italy, 2009; ISBN 978-88-470-1152-6.
39. Versteeg, H.K.; Malalasekera, W. *An Introduction to Computational Fluid Dynamics: The Finite Volume Method*; Pearson Education: London, UK, 2007; ISBN 978-0-13-127498-3.
40. Von Mises, R. Mechanik der festen Körper im plastisch- deformablen Zustand. *Nachr. Ges. Wiss. Gött. Math.-Phys. Kl.* **1913**, *1913*, 582–592.
41. Shaaban, A.M.; Duerinckx, A.J. Wall shear stress and early atherosclerosis: A review. *Am. J. Roentgenol.* **2000**, *174*, 1657–1665. [CrossRef] [PubMed]
42. Cheng, C.; Tempel, D.; van Haperen, R.; van der Baan, A.; Grosveld, F.; Daemen, M.J.A.P.; Krams, R.; de Crom, R. Atherosclerotic lesion size and vulnerability are determined by patterns of fluid shear stress. *Circulation* **2006**, *113*, 2744–2753. [CrossRef]
43. Hsiai, T.K.; Cho, S.K.; Honda, H.M.; Hama, S.; Navab, M.; Demer, L.L.; Ho, C.-M. Endothelial Cell Dynamics under Pulsating Flows: Significance of High Versus Low Shear Stress Slew Rates ($\partial\tau/\partial\tau$). *Ann. Biomed. Eng.* **2002**, *30*, 646–656. [CrossRef]
44. Millon, A.; Sigovan, M.; Boussel, L.; Mathevet, J.-L.; Louzier, V.; Paquet, C.; Geloen, A.; Provost, N.; Majd, Z.; Patsouris, D.; et al. Low WSS Induces Intimal Thickening, while Large WSS Variation and Inflammation Induce Medial Thinning, in an Animal Model of Atherosclerosis. *PLoS ONE* **2015**, *10*, e0141880. [CrossRef]
45. O'Leary, S.A.; Mulvihill, J.J.; Barrett, H.E.; Kavanagh, E.G.; Walsh, M.T.; McGloughlin, T.M.; Doyle, B.J. Determining the influence of calcification on the failure properties of abdominal aortic aneurysm (AAA) tissue. *J. Mech. Behav. Biomed. Mater.* **2015**, *42*, 154–167. [CrossRef]
46. Chatziprodromou, I.; Tricoli, A.; Poulikakos, D.; Ventikos, Y. Haemodynamics and wall remodelling of a growing cerebral aneurysm: A computational model. *J. Biomech.* **2007**, *40*, 412–426. [CrossRef] [PubMed]
47. Sheidaei, A.; Hunley, S.C.; Zeinali-Davarani, S.; Raguin, L.G.; Baek, S. Simulation of abdominal aortic aneurysm growth with updating hemodynamic loads using a realistic geometry. *Med. Eng. Phys.* **2011**, *33*, 80–88. [CrossRef]
48. Reed, D.; Reed, C.; Stemmermann, G.; Hayashi, T. Are aortic aneurysms caused by atherosclerosis? *Circulation* **1992**, *85*, 205–211. [CrossRef] [PubMed]
49. Boussel, L.; Rayz, V.; McCulloch, C.; Martin, A.; Acevedo-Bolton, G.; Lawton, M.; Higashida, R.; Smith, W.S.; Young, W.L.; Saloner, D. Aneurysm growth occurs at region of low wall shear stress: Patient-specific correlation of hemodynamics and growth in a longitudinal study. *Stroke* **2008**, *39*, 2997–3002. [CrossRef] [PubMed]

50. Xu, C.; Zarins, C.K.; Glagov, S. Aneurysmal and occlusive atherosclerosis of the human abdominal aorta. *J. Vasc. Surg.* **2001**, *33*, 91–96. [CrossRef] [PubMed]
51. Drewe, C.J.; Parker, L.P.; Kelsey, L.J.; Norman, P.E.; Powell, J.T.; Doyle, B.J. Haemodynamics and stresses in abdominal aortic aneurysms: A fluid-structure interaction study into the effect of proximal neck and iliac bifurcation angle. *J. Biomech.* **2017**, *60*, 150–156. [CrossRef]
52. De Heer, L.M.; Budde, R.P.J.; Mali, W.P.T.M.; de Vos, A.M.; van Herwerden, L.A.; Kluin, J. Aortic root dimension changes during systole and diastole: Evaluation with ECG-gated multidetector row computed tomography. *Int. J. Cardiovasc. Imaging* **2011**, *27*, 1195–1204. [CrossRef]

fluids

MDPI

Article

Comparative Study of PEGylated and Conventional Liposomes as Carriers for Shikonin †

Stella K. Tsermentseli, Konstantinos N. Kontogiannopoulos, Vassilios P. Papageorgiou and Andreana N. Assimopoulou *

Organic Chemistry Laboratory, School of Chemical Engineering, Aristotle University of Thessaloniki, 54124 Thessaloniki, Greece; s.tsermentseli@gmail.com (S.K.T.); kkontogiannopoulos@gmail.com (K.N.K.); vaspap@eng.auth.gr (V.P.P.)

* Correspondence: adreana@eng.auth.gr; Tel./Fax: +30-231-099-4242

† The authors would like to dedicate this work to the memory of Prof. Dr. Angelos Sagredos, who has recently passed away.

Received: 25 April 2018; Accepted: 25 May 2018; Published: 26 May 2018

Abstract: Liposomes are considered to be one of the most successful drug delivery systems. They apply nanotechnology to potentiate the therapeutic efficacy and reduce the toxicity of conventional medicines. Shikonin and alkannin, a pair of chiral natural naphthoquinone compounds, derived from *Alkanna* and *Lithospermum* species, are widely used due to their various pharmacological activities, mainly wound healing, antioxidant, anti-inflammatory and their recently established antitumor activity. The purpose of this study was to prepare conventional and PEGylated shikonin-loaded liposomal formulations and measure the effects of different lipids and polyethylene glycol (PEG) on parameters related to particle size distribution, the polydispersity index, the zeta potential, drug-loading efficiency and the stability of the prepared formulations. Three types of lipids were assessed (1,2-Dioleoyl-sn-glycero-3-phosphocholine (DOPC), 1,2-Distearoyl-sn-glycero-3-phosphocholine (DSPC) and 1,2-distearoyl-sn-glycero- 3-phospho-rac-(1-glycerol) (DSPG)), separately and in mixtures, forming anionic liposomes with good physicochemical characteristics, high entrapment efficiencies (varying from 56.5 to 89.4%), satisfactory in vitro release profiles and good physical stability. The addition of the negatively charged DSPG lipids to DOPC, led to an increment in the drug's incorporation efficiency and reduced the particle size distribution. Furthermore, the shikonin–loaded PEGylated sample with DOPC/DSPG, demonstrated the most satisfactory characteristics. These findings are considered promising and could be used for further design and improvement of such formulations.

Keywords: alkannin; cancer; stability study; drug delivery system

1. Introduction

Alkannin and shikonin (A/S; Figure 1), a chiral pair of natural naphthoquinone compounds, biosynthesized in the roots of more than 150 species of the Boraginaceae plant family (such as *lithospermum*, *Alkanna*, *Anchusa* and *Echium)* are widely used due to their various pharmacological activities. Biological investigations have established that A/S possess a wide spectrum of biological activities, such as strong wound healing, and tissue regenerative, anti-inflammatory, antioxidant and, most prominently, antitumor activity. It is worth-mentioning that alkannin and shikonin (the *S*- and *R*-isomer, respectively; Figure 1) display similar levels of pharmacological activity [1–6].

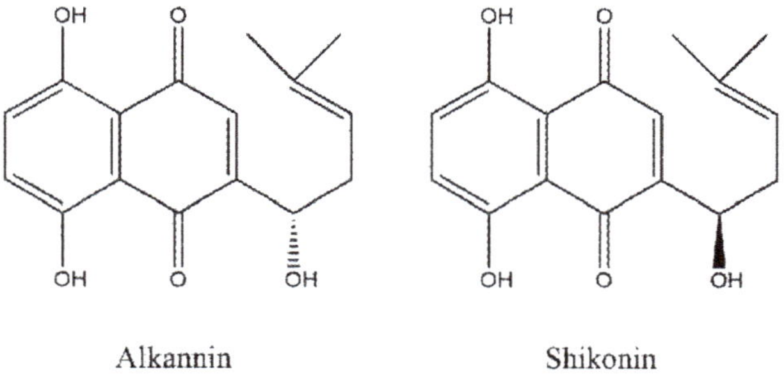

Figure 1. The enantiomers, alkannin and shikonin.

The great scientific interest and clinical potential for the antitumor properties of A/S in the development of novel chemotherapeutics and in effective combination chemotherapy is depicted by the large number of papers (more than 140) that have appeared in the literature the last five years [7,8]. Recent studies confirmed that shikonin has the potential to induce apoptosis in a variety of human tumor cell lines, including leukemia cell lines in vitro and in vivo, with minimal or no toxicity to healthy human cells [9–11].

This antitumor activity of shikonin may be attributed to its accumulation in the mitochondria of cancer cells, which disrupts mitochondrial function, and eventually causes apoptosis [12]. Our group investigated the inhibition of c-MYC expression and transcriptional activity by shikonin as a novel mechanism for killing leukemia cells [9], and more recently, the cytotoxic activity of shikonin to the Huh7 cancer cell line by a metabolite profiling approach which could set a basis for the elucidation of their antitumor activity was investigated [13]. Shikonin has also been proposed as a novel dietary agent with great potential in breast cancer prevention [14] and has been found to act synergistically to potentiate doxorubicin-induced growth inhibition and apoptosis in vitro [15].

These studies prove the potent antitumor activities of shikonin in multiple tumors through targeting multiple signaling pathways, promoting the necessity of solving the problem of drug resistance. Therefore, the most promising delivery systems for A/S derivatives need to be developed and optimized to exploit and assess their anticancer properties.

Nanoscale drug carriers offer the potential to improve the therapeutic index of drug molecules by diminishing their toxicity against physiological tissues. Furthermore, they can result in controlled therapeutic levels of the drug for a prolonged time. A proper drug delivery agent could modify the solubility and improve the stability of candidate drugs, and lead to an improved ADME profile (Absorption, bioDistribution, Metabolism, and Excretion) [16,17].

Liposomes represent an advanced type of nanotechnology that has the potential to target active molecules (anticancer agents, peptide hormones, enzymes, proteins, vaccines) to the site of action, improving the therapeutic index [18–20]. Many clinical studies have shown that liposomes have improved the pharmacokinetics and biodistribution of therapeutic agents. Several anticancer liposomal formulations (conventional and PEGylated) have been approved and are commercially available, such as DOXIL®/Caelyx® (doxorubicin), Lipo-Dox® (doxorubicin), Myocet®/Evacet® (doxorubicin), DaunoXome® (daunorubicin), Myocet®/Evacet® (doxorubicin), Ambisome® (Amphotericin B) and Marqibo® (Vincristine) [21,22].

A significant improvement came with the incorporation of PEG-lipid leading liposomes (PEGylated or Stealth® liposomes) which remain for longer time periods in the blood circulation. The presence of PEGs on the surface of liposomes prevents their uptake by the reticuloendothelial system (RES), which is attributed to their highly hydrated surfaces due to the hydrophilic polymers that result in the inhibition of protein adsorption and opsonization of the liposomes [23]. In this way, liposomes have, to some extent, the ability to pass in and out of the liver and spleen, avoiding clearance, and thus remain in the tumor tissue due to the depleted lymphatic drainage [24,25]. In previous studies, anticancer agents that were incorporated in PEGylated liposomes displayed longer circulation times and enhanced drug delivery to tumor tissues [26].

There is great interest in exploiting the wide range of anticancer activities of the hydrophobic A/S and derivatives towards several tumors, and therefore an optimum administration system needs to be developed. The incorporation of a hydrophobic drug (such as shikonin) into liposomes improves its bioavailability and leads to increased stability and anticancer activity, along with decreased drug toxicity. Regarding shikonin, it has been recently proven that liposomes significantly decrease its toxicity in vitro and in vivo [27]. Furthermore, under the frame of developing drug delivery systems with alkannins and shikonins as bioactive molecules, we have successfully incorporated shikonin into both conventional (with lipids such as egg phosphatidylcholine (EPC), 1,2-dipalmitoylphosphatidylcholine (DPPC) and 1,2-distearoyl-sn-glycero-3-phosphocholine (DSPC)) [28] and stealth liposomes (DSPC-PEG$_{2000}$, EPC-PEG$_{2000}$, DPPC-PEG$_{2000}$) [29].

In this regard, the scope of this paper was to expand our previous research by preparing and characterizing shikonin-loaded liposomes, with different types of lipids, aiming to produce an optimized formulation and to compare this with the ones already prepared. Specifically, DOPC (1,2-dioleoyl-sn-glycero-3-phosphocholine) and a mixture of lipids with the charged lipid DSPG (1,2-distearoyl-sn-glycero-3-phospho-rac-(1-glycerol) sodium salt), like DOPC/DSPG and DSPC/DOPC, were used for the first time, for both conventional and stealth liposomes. The negatively charged lipid was used to prepare anionic liposomal formulations in order to prevent the aggregation of liposomes due to electrostatic repulsion [24,30,31]. Furthermore, it was reported that the presence of negatively charged lipids in liposomes, allows rapid uptake by the reticuloendothelial system [24], while large and positively charged liposomes induce cytokine activation and toxicity and thereby their safety for clinical use is limited [32].

Thus, three conventional and PEGylated liposomal formulations of shikonin (DOPC, DOPC/DSPG and DSPC/DSPG) were formulated and characterized with consideration of their physicochemical characteristics (particle size distribution, ζ-potential, entrapment efficiency), in vitro release profiles and physicochemical stability (4 °C for a 28 days period: drug leakage, particle size distribution, ζ-potential). The new formulations were also compared with our previously reported shikonin-loaded liposomes [28,29].

2. Materials and Methods

2.1. Materials

Shikonin was purchased from Ichimaru Pharcos Co., Ltd. (Gifu, Japan) and was used after purification through column chromatography followed by recrystallization with n-hexane, in accordance with Assimopoulou et al. [33] (purity obtained: 100% by HPLC-DAD, Agilent Technologies, Waldbronn, Germany).

1,2-distearoyl-sn-glycero-3-phosphocholine (DSPC; MW 790.15); 1,2-dioleoyl-sn-glycero-3-phosphocholine (DOPC; MW 786.11); and 1,2-distearoyl-sn-glycero- 3-phospho-rac-(1-glycerol) sodium salt) (DSPG; MW 801.06) were generously donated by Lipoid GmbH (Ludwigshafen, Germany). *N*-(Carbonyl-methoxypolyethyleneglycol 2000)—1,2 distearoyl-en-glycero-3-phosphoethanolamine (DSPE-mPEG2000; MW 2806.0) was purchased from Genzyme Pharmaceuticals (Cambridge, MA, USA). Phosphate buffer saline of pH 7.4 (PBS), cholesterol (CHOL), sodium lauryl sulfate (SLS), Sephadex G75 and dialysis sacks (molecular weight cut off 13,000) were obtained from Sigma–Aldrich (St. Louis, MO, USA). Organic solvents used for all experiments were of analytical grade and were purchased from Sigma–Aldrich (St. Louis, MO, USA), and water was of HPLC grade.

2.2. *Liposome Preparation*

Shikonin-loaded liposomes were formulated using the thin-film hydration method. In brief, lipids, cholesterol and shikonin were dissolved in $CHCl_3$/MeOH 2:1 (v/v) using the same molar ratios for all samples: lipid/CHOL (4.5:1), neutral lipid/charged lipid (9:1), lipid/DPSE-mPEG$_{2000}$ (13:1) and lipid/shikonin (30:1) (see Table 1). Organic solvent was slowly removed in a rotary evaporator (EYELA N-N Series, Digital Water bath SB–651, Tokyo, Japan), forming a thin lipid film on the flask. Solvent traces were removed by leaving the flask overnight under vacuum. The lipid film was then hydrated by the addition of PBS (6.5 mL for drug-loaded liposomes and 2 mL for liposomes without drugs—"empty liposomes") for 1.5 h in a water bath. The temperature was maintained above the main phase transition temperature (T_m) of each lipid (−20 °C for DOPC, 67 °C for DSPC and 55 °C for DSPG). Flask was vortexed in an IKA MS2 Minishaker (IKA Works Inc., Wilmington, NC, USA) at 1500 rpm for 10 min.

Table 1. Liposome compositions used in the study.

Sample	Type of Lipids	Lipid (mg) [a]	DSPE-mPEG$_{2000}$ (mg) [b]	Cholesterol (mg) [c]	Shikonin (mg) [d]
1	DOPC	120	-	13.2	1.47
1e	DOPC	30	-	3.27	-
2	DOPC/DSPG	120	-	13.07	1.46
2e	DOPC/DSPG	30	-	3.27	-
3	DSPC/DSPG	120	-	13.01	1.09
3e	DSPC/DSPG	30	-	3.25	-
4	DOPC	120	33.12	13.2	1.47
4e	DOPC	30	8.28	3.27	-
5	DOPC/DSPG	120	33.06	13.07	1.46
5e	DOPC/DSPG	30	8.26	3.27	-
6	DSPC/DSPG	120	32.91	13.01	1.09
6e	DSPC/DSPG	30	8.23	3.25	-

All the samples were prepared in triplicate. e stands for "empty" formulations (drug-free). [a] The neutral lipid/charged lipid molar ratio was 9:1 for samples containing DSPG (1,2-distearoyl-sn-glycero-3-phospho-rac-(1-glycerol) sodium salt). [b] The lipid (or mixture of lipids)/DPSE-mPEG$_{2000}$ (distearoyl-en-glycero-3-phosphoethanolamine) molar ratio was 13:1 for PEGylated samples. [c] The lipid (or mixture of lipids)/cholesterol (CHOL molar ratio was 4.5:1 for all samples. [d] Lipid (or mixture of lipids)/drug molar ratio was 30:1 for all samples. DOPC: 1,2-dioleoyl-sn-glycero-3-phosphocholine.

Small unilamellar vesicles (SUVs) were obtained from the resultant multilamellar vesicles (MLVs) by probe sonication, using a Sonicator W–375 Cell Disruptor (Heat Systems–Ultrasonics Inc.) for 2×5 min periods, interrupted by a 5 min rest period in ice bath (amplitude 0.6; pulser 50%). Formulations were left for 30 min to anneal any structural defects.

2.3. *Characterization of Shikonin-Loaded Liposomes*

2.3.1. Particle Size and ζ-Potential

The size distribution and ζ-potential of the liposomal formulations were measured immediately after preparation by dynamic light scattering using a Malvern ZetaSizer Nano ZS (Malvern Instruments, Malvern, UK) at 25 °C. Prior to measurement, all samples were diluted by 60-fold in PBS (pH 7.4).

2.3.2. Entrapment Efficiency

An aliquot of freshly prepared, loaded liposomes (300 µL) was transferred to a size exclusion column (Sephadex G75) and eluted with PBS in order to separate free from entrapped shikonin and to determine the entrapment efficiency. Purified liposomes (500 µL) were diluted in methanol (2.5 mL) to destroy liposomal structure and release the drug into the organic phase. The concentration of shikonin was determined by the absorbance of the organic phase (measured with a Ultraviolet–visible spectroscopy (UV–Vis) Hitachi U1900, Hitachi High-Technologies Corporation, Tokyo, Japan) using the following calibration curve:

$$\text{Shikonin Concentration (mg/mL)} = 0.0316 \times \text{Absorbance} - 0.00009; (R^2 = 0.9999). \quad (1)$$

The entrapment efficiency was calculated as follows:

$$\text{Entrapment Efficiency (\%)} = (F_i/F_t) \times 100, \quad (2)$$

where F_i is the shikonin concentration into liposomes and F_t is the initial concentration of shikonin.

2.3.3. In Vitro Release

Briefly, 3 mL of shikonin-loaded liposomes were inserted into dialysis sacks (molecular weight cut off 13,000; Sigma–Aldrich). The sealed dialysis sacks were incubated in 20 mL of release medium (PBS + 1% SLS) at 37 °C in a water bath and stirred magnetically (RET control-visc, IKA Werke, Germany). Aliquots of release medium (3 mL) were withdrawn at specific time intervals, to determine the accumulative amount of drug released, and they were replaced with fresh release medium. The shikonin concentration in the release medium was calculated with a UV–Vis spectrometer at λ_{max} = 516 nm with the aid of the following calibration curve (shikonin in release medium):

$$\text{Shikonin Concentration (mg/mL)} = 0.0538 \times \text{Absorbance} - 0.0002; (R^2 = 0.9999). \quad (3)$$

Release curves were drawn according to the cumulative drug release and plotted vs time (with the aid of the following equation):

$$\text{\% Cumulative Shikonin Released}_t = \text{Shikonin Released}_t/\text{Total Entrapped Shikonin} \times 100. \quad (4)$$

2.4. *Stability*

Immediately after their preparation, samples were stored in dark glass vials (in their hydrated form) at 4 °C for 28 days, in order to study their stability. Aliquots were withdrawn at specific time intervals and assessed in terms of their mean particle size, ζ-potential and drug retention.

2.5. *Statistical Analysis*

All results are expressed as mean values ± standard deviations of three independent experiments. Statistical analysis was performed using SPSS 22.0 software (SPSS Inc., Chicago, IL, USA). In order to examine the statistical significance between samples, multiple comparisons were performed by one-way analysis of variance (ANOVA) followed by post-hoc analysis using Tukey's test. $p < 0.05$ was considered statistically significant.

3. Results and Discussion

In the present study, three lipid types (DOPC, DOPC/DSPG, and DSPC/DSPG) were utilized to prepare conventional and PEGylated liposomes containing shikonin, aiming to reduce the drug's toxicity and achieve controlled release. This is a continuation of the current authors' active research on developing and evaluating drug delivery systems for bioactive naphthoquinones, such as alkannins and shikonins [28,29].

3.1. *Liposome Characterization*

The physicochemical characterization of liposomes, such as their size, shape and charge are vital parameters in the delivery of improved bio-distribution and prolonged pharmacokinetics of encapsulated cytotoxic drugs [34]. The prepared liposomal formulations (shikonin-loaded and without the drug) were characterized in terms of their particle size distribution and ζ-potential values. Furthermore, the amount of drug incorporated and the release kinetics of the drug were additionally estimated (Table 2).

Table 2. Physicochemical characteristics of liposomal formulations.

Sample	Mean Particle Size (nm)	Polydispersity Index (PDI)	ζ-Potential (mV)	Entrapment Efficiency (%)
1	113.62 ± 9.72	0.29 ± 0.02	−6.10 ± 1.11	66.67 ± 0.04
1e	79.15 ± 1.34	0.31 ± 0.01	−11.41 ± 2.68	—
2	84.90 ± 0.10	0.25 ± 0.03	−19.88 ± 4.31	78.42 ± 0.01
2e	85.95 ± 1.20	0.29 ± 0.07	−26.97 ± 1.30	—
3	222.0 ± 6.56	0.33 ± 0.05	−16.23 ± 9.29	56.50 ± 0.03
3e	111.67 ± 6.56	0.25 ± 0.02	−13.33 ± 2.01	—
4	81.92 ± 0.11	0.18 ± 0.02	−14.62 ± 2.47	75.64 ± 0.04
4e	62.33 ± 4.22	0.19 ± 0.01	−27.93 ± 7.74	—
5	70.42 ± 1.25	0.17 ± 0.02	−16.68 ± 3.61	89.40 ± 0.02
5e	71.15 ± 0.21	0.20 ± 0.01	−19.17 ± 1.07	—
6	121.33 ± 3.77	0.29 ± 0.01	−13.38 ± 0.49	66.85 ± 0.01
6e	93.00 ± 1.00	0.23 ± 0.03	−11.54 ± 1.54	—

3.1.1. Particle Size Measurement

The size of the liposomes affects their residence time in the systemic blood circulation, as well as their pathways within the body. The smaller the size, the more difficult it is for them to become detectable by the macrophages of the immune system, increasing in this way the residence time in the systemic circulation, as well as their effectiveness [35,36].

Regarding conventional formulations (samples 1, 2 and 3), the mean particle size varied from 84.9 nm to 222 nm, while for the corresponding PEGylated liposomes (samples 4, 5, and 6), sizes ranged from 70.4 nm to 121.3 nm (Table 2), giving PEGylated liposomes a significant advantage.

The lipid type significantly affected the mean particle size of the liposomal formulations. As previously reported, unsaturated fatty acids can incur oxidative reactions, altering the permeability of the liposomal bilayers [37] and if the degree of unsaturation of the fatty acid side chains is too high, there is a possibility that stable liposomes might not be formed [38]. In another study, it was reported that the average liposome size varies between 30 and 300 nm, depending on the lipid composition and ionic strength of the lipid mixture during liposome formation. Furthermore, the size effect of the charged lipids should be a factor in choosing a lipid mixture [39]. Additionally, in our previous research, it was noticed that mean particle size was affected by the lipid type as well as by the interactions between the lipid bilayer and shikonin [29].

Shikonin-loaded DOPC conventional liposomes (sample 1) presented a statistically smaller mean particle size (113.6 nm) compared to DSPC and DPPC (221.2 and 243.2 nm, respectively) and similar to EPC (144.5 nm) which were measured in our previous work [29]. The addition of a negatively charged lipid (DSPG) to DOPC (sample 2) led to a conventional formulation with a smaller mean particle size (84.9 nm) and lower polydispersity index (0.25). On the other hand, the addition of DSPG to DSPC did not significantly affect the particle size (222 nm) or polydispersity index (0.33) of DSPC formulations, resulting in liposomes with larger mean particle size compared to DOPC/DSPG ones. The addition of DSPG had different impacts on particle size distribution, depending on the type of lipid that it was combined with. Similar results arose from the corresponding formulations without the drug (samples 1e, 2e and 3e), as well as from other published studies [40]. Furthermore, it was reported that the ionic strength of the mixture during the formation of liposomes might influence their mean diameter in the presence of charged lipid components. Moreover, the ability to control the average liposome size is by varying the proportion of charged lipid components [39].

The use of DSPE-mPEG$_{2000}$ resulted in PEGylated formulations (samples 4, 5, and 6) with smaller mean particle sizes and lower polydispersity indexes (Figure 2; the reduction varied between 17%–45% for drug-loaded liposomes and 17%–21% for drug-free liposomes). Briefly, PEGylated shikonin-loaded liposomes with DOPC (sample 4, 81.9 nm) had statistically smaller sizes in comparison to those in

DSPC formulations (124.8 nm), and similar sizes to the EPC and DPPC liposomes (93 and 105.9 nm respectively) produced in our previous work [29]. Combining DSPG with DOPC and DSPC lipids led to PEGylated formulations with similar (sample 5; 70.4 nm) or statistically lower (sample 6; 121.3 nm) mean particle sizes compared to the conventional samples (sample 2 and 3 respectively). These results agree with the observations from the PEGylated drug-free formulations (samples 4e, 5e and 6e) as well as with similar papers in the literature [40], confirming the superiority of PEGylated formulations over conventional ones.

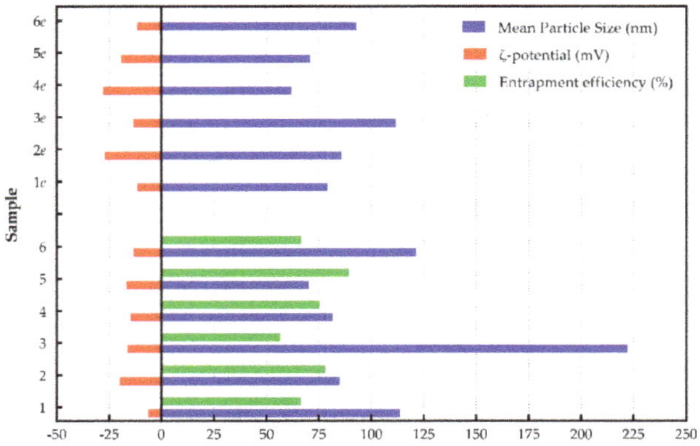

Figure 2. Comparison of liposomal formulations.

This decrease observed in particle size could be explained by the curving of the bilayer to reduce the intensity of lateral repulsion, caused by the addition of increasing amounts of PEG in the lipid bilayer. PEGylated lipid also increases interlamellar repulsion, causing a decrease in lamellarity [41]. A large number of studies with other bioactive molecules have confirmed this trend [31,41,42].

3.1.2. ζ-Potential

The ζ-potential is the electrostatic charge of the particle surface which acts as a repulsive energy barrier controlling the stability of dispersion and opposing the aggregation of liposomes in buffer solution [43]. The charge of the phospholipids could affect the pharmacodynamic and pharmacokinetic properties of the liposomes, as well as tumor accumulation. For instance, cationic liposomes are more readily taken up by cells in comparison with anionic or neutral liposomes, due to attractive forces between the positively charged liposomes and the negatively charged outer cell membrane. However, such interactions may also damage the cell membrane, resulting in toxicity, and they have been found to cause pulmonary toxicity, due to the generation of reactive oxygen species [44].

The ζ-potential values of all samples were measured immediately after preparation. As indicated in Table 2, the ζ-potential values of conventional shikonin-loaded liposomal formulations (samples 1, 2 and 3) varied from −6.1 mV to −19.9 mV, while the ζ-potentials of the corresponding PEGylated liposomes (samples 4, 5, and 6) ranged from −13.4 mV to −16.7 mV. Similar values for other bioactive constituents have been reported in the literature [29,45,46].

Liposome's ζ-potential values indirectly reflect the net charge of the vesicle surface. This is why the type of lipid used significantly influences the charge of the liposomal formulation [47]. Figure 2 depicts that the use of different type of lipids resulted in liposomal formulations with significant differences in ζ-potential values. More precisely, the use of DOPC lipid formed conventional liposomes (sample 1; -6.1 mV) with statistically lower ζ-potential values compared to EPC (-16.6 mV) or similar ζ-potential values compared to DSPC and DPPC lipids (-7.3 and -8.4 mV respectively) [29]. This may be attributed to interactions taking place between shikonin and the type of lipid. The addition of DSPG to DOPC and

DSPC lipids formed shikonin-loaded conventional liposomes with statistically increased ζ-potential values (-19.9 and -16.2 mV, respectively). This could be due to the fact that DSPG is a negatively charged lipid, adding an additional charge in the lipid bilayer [40,48]. In addition, DOPC/DSPG liposomes (sample 2) seemed to have an advantage regarding their ζ-potential values, compared to DSPC/DSPG (sample 3). Furthermore, although in our previous work, "empty" liposomes presented decreased ζ-potential values compared to shikonin-loaded ones [29], in the present study, DOPC lipids led to the opposite trend. This trend could be attributed to the presence of shikonin, resulting in a decrease of ζ-potential values, since drug incorporation causes significant variations in the liposomal surface structure and the orientation of the phosphatidylcholine head group [49].

All drug-loaded PEGylated formulations (samples 4, 5, and 6) showed statistically increased ζ-potential values compared to the corresponding conventional liposomes (samples 1, 2 and 3). Liposomes formed with DOPC (sample 4) exhibited a ζ-potential value of -14.6 mV, which is similar to formulations previously prepared from our group with DSPC and EPC [29]. Moreover, the use of DSPG mixed with DOPC or DSPC, did not provoke any significant changes in the ζ-potential of the prepared liposomes. The same trends were obtained for the corresponding "empty" liposomes. Formulations with DSPE-PEG_{2000} have less negative charge compared to liposomes without DSPE-PEG_{2000}, a fact attributed to the "masking" of some of the anionic charges of DSPG by DSPE-PEG_{2000}, which could explain the observed trend [40,47,50].

3.1.3. Entrapment Efficiency

The entrapping efficiency of a drug into the liposome structure depends on many factors, such as the ionic strength of the buffer, the pH, the incubation time, the drug loading ratio, the lipid composition and others [51]. For the shikonin-loaded conventional liposomes (samples 1–3), the entrapment efficiency varied from 56.5 to 78.4% (as shown in Table 2), while the values of the corresponding PEGylated ones (samples 4–6) ranged from 66.9 to 89.4%.

The results of the current study showed significant variation in entrapment efficiency among samples prepared with different lipids. More precisely, the use of DOPC lipids resulted in a significant lower entrapment efficiency (66.7%) compared to similar DSPC (85.3%) and DPPC (77.9%) formulations and a higher efficiency than EPC liposomes (52.9%) [29]. The influence of the charged DSPG depended on the type of lipid that it was mixed with, improving DOPC's performance (sample 2; 78.4%) and diminishing DSPC's performance (sample 3; 56.5%). These results are in accordance with other published studies [47,52], and it was observed that encapsulation was affected by electrostatic interactions between the drug's peripheral surface and the polar head groups of phospholipids. In addition, there was a linear correlation between the lipid concentration and the encapsulation efficiency [53,54].

Regardless of the type of lipid used, all PEGylated formulations (samples 4, 5 and 6) had increased drug incorporation compared to conventional ones (this increase varied from 13 to 18%). This could be due to the presence of PEG, placed on the outer surface of the lipid bilayer, causing an increase in drug entrapment within the bilayer [54–56]. More specifically, PEGylated liposomes with DOPC lipids showed a similar entrapment efficiency (75.6%) to previously prepared PEGylated formulations with DSPC, EPC and DPPC lipids (91.8%, 71.8%, and 84.9% respectively) [29]. These observations are in accordance with similar studies [50] and imply that shikonin incorporation could be strengthened by the presence of PEG on the outer surface of the lipid bilayer.

3.1.4. In Vitro Drug Release

The drug release profiles of all shikonin-loaded formulations were assessed (as depicted in Figure 3). In addition, we calculated the amount (%) of shikonin released, in regard to the total entrapped drug, after 8 and 72 h, as well as the required time period to release 50% of the total released drug ($t_{50\%}$) (as shown in Table 3).

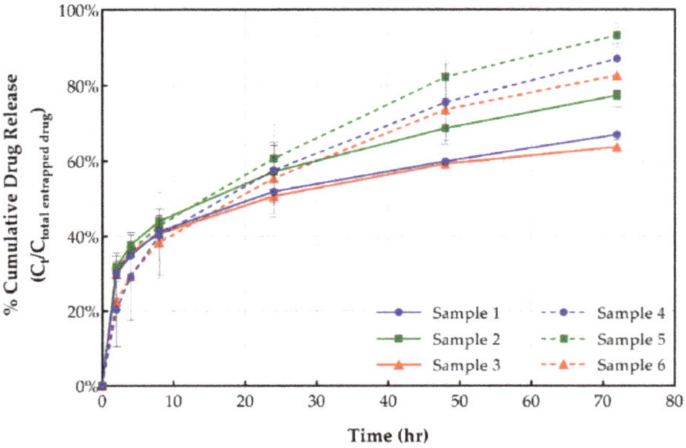

Figure 3. Cumulative release from PEGylated and conventional liposomes vs time.

Table 3. Release data of shikonin-loaded formulations.

Sample	Total Drug Release (72 h) (%)	Drug Release at 8 h (%)	$t_{50\%}$ (h)
1	67.07 ± 0.01	41.24 ± 0.03	3.86
2	77.41 ± 0.01	43.95 ± 0.01	7.04
3	63.80 ± 0.03	40.67 ± 0.02	3.57
4	87.0 ± 0.08	40.22 ± 0.11	18.26
5	93.25 ± 0.03	42.76 ± 0.00	18.45
6	82.48 ± 0.08	38.36 ± 0.09	17.93

As depicted in Table 3, PEGylated liposomes (samples 4–6) released approximately 20%–30% more shikonin compared to the corresponding conventional formulations (samples 1–3). In addition, Figure 3 depicts an interaction between the shikonin release rate and the type of lipid used. Briefly, samples with DOPC lipid released a statistically greater amount of drug (87%) after 72 h, in comparison to similar, previously reported shikonin-loaded liposomal formulations (60.1% for samples with EPC, 63.3% for samples with DSPC and 66.9% for samples with DPPC) [29]. The addition of DSPG seems to improve the release profile of all formulations, increasing the amount of drug released over a 72 h period.

Conventional samples, on the other hand, appear to follow the same trend, with DOPC liposomes releasing a greater amount of shikonin (67.1%) compared to other formulations with EPC (20.9% release), DSPC (51.4% release) and DPPC (47.5% release) prepared by Kontogiannopoulos et. al. [29]. Moreover, DSPG increased the amount of drug released from conventional liposomes, as well.

As shown in Table 3, the type of lipid used also influenced the required time period to release 50% of the total released drug ($t_{50\%}$). Briefly, the $t_{50\%}$ for conventional samples ranged between 3.6 and 7.0 h and between 17.9 and 18.5 h for the corresponding PEGylated liposomes. In both cases, the DOPC:DSPG mixture exhibited the higher value. These observations, are in line with the report that longer alkyl chain lipids enhance the binding of the drug with the lipid bilayer, resulting in slower or sustained drug release [57].

The different behaviour in terms of the release rate between the PEGylated and conventional liposomes may be attributed to the bilayer rigidity—the more rigid the bilayer, the slower the release of the drug [50]. Thus, it can be assumed that shikonin's release profile could be modified by the existence of PEG. Furthermore, differences among the prepared formulations could be attributed to several factors, such as lipid type and dose, interactions between the drug and lipid bilayer, as well as the lipid–PEG conjugate in the case of stealth liposomes [58]. Each type of phospholipid affects the efflux rate in a different way, e.g., a higher degree of saturation and an increased fatty acid chain-length, retarding the leakage rate of molecules from liposomes [37].

3.2. Stability Study

The stability of any pharmaceutical formulation is essential. Thus, all samples were maintained in their hydrated state for 28 days at 4 °C, while drug leakage, mean particle size, the polydispersity index (PDI) and the ζ-potential were monitored in order to assess their stability.

3.2.1. Particle Size Distribution

The results from the stability study depicted a significant variability in the mean particle size and PDI in the conventional shikonin-loaded formulations. On the contrary, PEGylated liposomes remained more stable after 28 days of preservation at 4 °C (Figure 4). Regarding conventional liposomes, sample 1 (prepared with DOPC) presented a size increment of 80.2%, which was higher than similar DSPC samples have shown (52.3% in particular) and lower than DPPC samples (104% increment) [29]. Moreover, when DSPG was added to DOPC (sample 2) or DSPC (sample 3) led to the formation of more stable conventional formulations (the observed size increases were 27.6% and 41.9%, respectively).

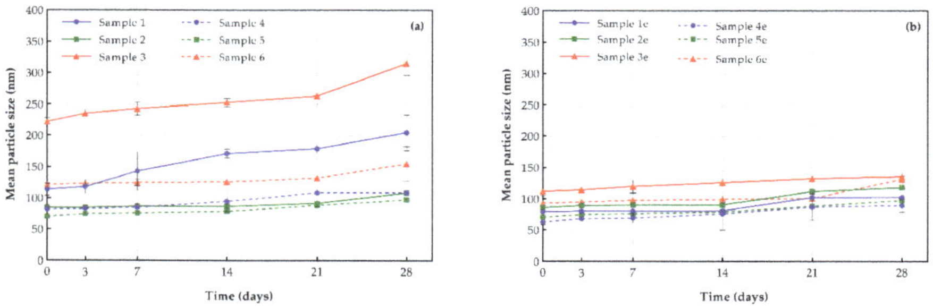

Figure 4. Mean particle size stability of (**a**) drug-loaded and (**b**) drug-free liposomal formulations (storage conditions: 4 °C, 28 days).

Concerning the prepared PEGylated formulations, liposomes with DOPC (sample 4) were less stable (32.8% increment) than contiguous samples of DSPC, EPC and DPPC (increases of 9.5%, 2.5%, and 6%, respectively) from our previous work [29]. It is perceived that the addition of DSPG to DOPC (sample 5) or DSPC (sample 6) lipids, led to formulations with similar rates of increase in particle size (38.4% and 41.9%, respectively). It is worth mentioning that conventional shikonin-loaded liposomes prepared with DOPC/DSPG remained particularly stable in terms of their particle size distribution and PDI, with a final mean particle size (after their residence for 28 days at 4 °C) that was close enough to PEGylated systems.

Both PEGylated and conventional formulations appeared to be particularly stable until day 21, and a slight increment in their mean size was observed in the period from day 21 to day 28 (Figure 4). Finally, it was observed that the increased rate in mean particle size of "empty" DOPC/DSPG (sample 2e) and DOPC/DSPG-PEG (sample 5e) liposomes was slightly higher (38.3% and 59.5% respectively) compared to the corresponding drug-loaded systems (samples 2 and 5; 27.6% and 38.4% respectively).

These observations confirm the supremacy of PEGylated liposomes regarding their particle size distribution stability over conventional formulations.

3.2.2. ζ-Potential

As depicted in Figure 5, the ζ-potential values of most samples remained stable (or reduced slightly) after their preservation at 4 °C. This indicates that the prepared samples maintained their initial charge, and as a result, their tendency for aggregation and flocculation was decreased. More precisely, DOPC conventional liposomes showed a lower rate of ζ-potential decrease (3.6%) in comparison to DSPC, EPC and DPPC (10.4%, 43.1%, and 30.5%, respectively) [29]. The addition of DSPG to DOPC and DSPC lipids led to similar rates of decrease (7.3% and 3.4%, respectively). Liposomes with DSPC/DSPG (sample 3) showed slight decreases in their ζ-potential values until day 7 and after that period, their ζ-potential values remained stable. On the other hand, DOPC/DSPG samples did not show any significant reduction during their residence at 4 °C for 28 days. Corresponding drug-free samples seemed to follow the same trend.

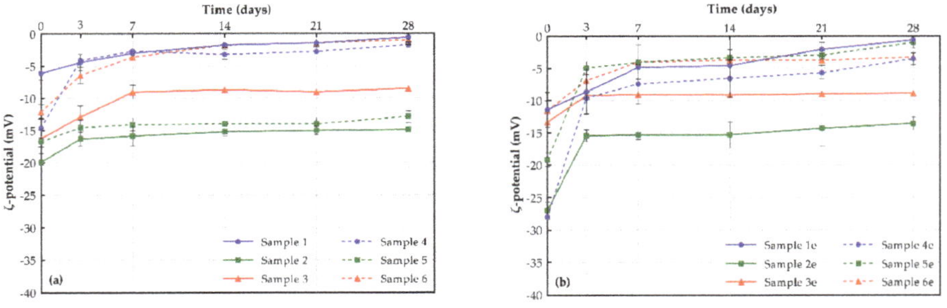

Figure 5. ζ-potential values of (**a**) drug-loaded and (**b**) drug-free liposomal formulations (storage conditions: 4°C, 28 days).

PEGylated shikonin-loaded liposomes (samples 4, 5, and 6) showed similar ζ-potential reduction rates to the conventional ones (samples 1, 2 and 3). Briefly, formulations with DOPC lipids (sample 4) showed lower rates of ζ-potential decrease (15.4%), compared to DSPC, EPC and DPPC lipids (26.8%, 28.5%, and 29%, respectively [29]. Their ζ-potential values only decreased until day 3 and from then on, they remained stable. The same results arose when DSPG was added either to DOPC or DSPC lipids, with PEGylated liposomes showing similar ζ-potential reduction rates (5.6% and 9.5%, respectively).

The study of the corresponding "empty" formulations resulted in similar observations. However, the DOPC/DSPG lipid mixtures (samples 2e and 5e) showed different behavior. More precisely, their ζ-potential values increased after day 3, in contrast with the corresponding drug-loaded samples (2 and 5) where the ζ-potential values remained almost stable across the entire 28-day period. This may be attributed to the presence of shikonin that helps this specific lipid mixture to be more stable, maintaining low ζ-potential values and retaining the initial charge, and therefore, limiting the tendency for agglomeration and flocculation.

3.2.3. Drug Leakage

Since, drug leakage is a crucial parameter for drug delivery systems, all samples were assessed for it over the 28-day stability period at 4 °C.

As depicted in Figure 6, all conventional shikonin-loaded liposomes (samples 1, 2 and 3) remained stable for 28 days, retaining more than 67.2% of their initially incorporated shikonin. More specifically, formulations prepared with DOPC lipid (sample 1) retained 74.3% of the initially incorporated drug, a significantly higher percentage compared to liposomes formed with DPPC lipid (62.3%) and a bit lower than those prepared with DSPC and EPC (81.4% and 83.1%, respectively) [29]. On the other hand, PEGylated formulations exhibited a higher amount of retention after 28 days, compared to all conventional samples, retaining, on average, ~85% of the initially incorporated drug. More precisely, the PEGylated liposomal formulation with DOPC lipids (sample 4) retained 86.5% of its initial drug content, almost the same as DSPC and EPC shikonin-loaded liposomes (85.2% and 86.4%) and a higher percentage of retention than DPPC (83.7%) [29]. In cases where DSPG was added to DOPC, there was an increment in drug stability for both conventional and PEGylated samples. On the contrary, if DSPG was mixed with DSPC (samples 3 and 6), the result was a higher drug leakage from the prepared liposomes.

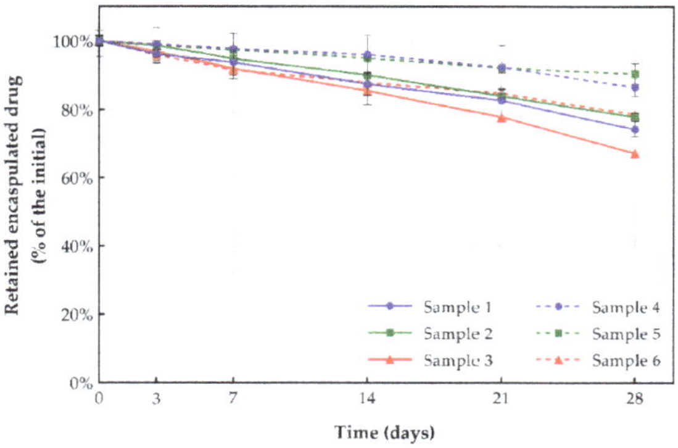

Figure 6. Drug retention of drug-loaded liposomal formulations (storage conditions: 4 °C, 28 days).

4. Conclusions

Liposomes represent an advanced nanoscale drug delivery system that could deliver bioactive constituents to a specific site of action. Alkannin, shikonin and their derivatives have proved to be potent chemotherapeutic and chemo-preventive agents that should be further exploited as novel chemotherapeutics and for effective combination chemotherapy. Therefore, drug delivery systems for shikonin need to be further developed and optimized to provide effective and safe administration and this paper is a continuation of the research of our group in this direction.

Shikonin-loaded conventional and PEGylated liposomes were successfully prepared using three types of lipid (DOPC, DOPC/DSPG, and DSPC/DSPG). All samples were assessed for their physicochemical characteristics, entrapment efficiency, drug release and stability during residence at 4 °C for 28 days.

All shikonin-loaded liposomes showed desirable drug entrapment efficiencies, varying from 66.9 to 89.4% for PEGylated types, to 56.5 to 78.4% for conventional systems, regardless of the type of lipid used. The use of negatively charged lipid (DSPG) combined with DOPC, increased incorporation efficiency values and helped to form liposomes with reduced particle sizes. Furthermore, both conventional and PEGylated shikonin-loaded liposome with DOPC/DSPG lipids remained particularly stable in terms of their particle size distribution and ζ-potential. Even after 28 days of residence at 4 °C, PEGylated shikonin-loaded formulations preserved more than 85% (on average) of the initially incorporated drug.

The presence of DSPE-mPEG$_{2000}$ in the outer surface of the lipid bilayer significantly modified the characteristics of formulations. More specifically, DSPE-mPEG$_{2000}$ caused a decrease in the mean particle size, regardless of the type of lipid used (reduction varied between 17%–45% for drug-loaded liposomes and 17%–21% for drug-free liposomes). The minimum liposome size was obtained with DOPC/DSPG lipids. Furthermore, PEGylated samples showed higher entrapment efficiencies compared to conventional samples (increments varied from 13 to 18%) and higher values were obtained with DOPC/DSPG lipids. Concerning drug release profiles, PEGylated formulations released approximately 20%–30% more shikonin over a prolonged time period, compared to the corresponding conventional formulations.

In conclusion, PEGylated formulations appear to be more advantageous than conventional ones and should be further exploited to increase the therapeutic index of shikonin. The addition of the charged lipid, DSPG, improved, in most cases, the physicochemical and pharmaceutical characteristics of liposomes. Conventional shikonin-loaded liposomes prepared with DOPC/DSPG showed the most promising results and stability (in terms of almost all parameters) compared to all other conventional shikonin-loaded formulations studied so far, with characteristics close enough to PEGylated systems. Respectively, the shikonin-loaded PEGylated sample with DOPC/DSPG lipids, showed the most satisfactory characteristics among all studied samples so far from our group. However, such systems are much too complicated and should be further examined in terms of their physicochemical interactions.

Author Contributions: Vassilios P. Papageorgiou and Andreana N. Assimopoulou had the initial conception of this work; Konstantinos N. Kontogiannopoulos designed the experiments; Stella K. Tsermentseli performed the experiments and acquired the data; Konstantinos N. Kontogiannopoulos analyzed the data, performed the visualization and wrote the original draft; Andreana N. Assimopoulou and Vassilios P. Papageorgiou had the overall supervision and project administration as well as they revised the manuscript. This work was part of the Ph.D. thesis of S. K. Tsermentseli.

Acknowledgments: The authors would like to thank Lipoid Company (Lipoid GmbH, Ludwigshafen, Germany) for generously donating 1,2-dioleoyl-sn-glycero-3-phosphocholine (DOPC), 1,2-distearoyl-sn-glycero-3-phosphocholine (DSPC) and 1,2-distearoyl-sn-glycero-3-phospho-rac-(1-glycerol) sodium salt) (DSPG). The authors are grateful to Em. Professor Constantinos Kiparissides (School of Chemical Engineering, AUTh) and Associate Professor Dimitrios Fatouros (School of Pharmacy, AUTh) for giving open access to their laboratories.

Conflicts of Interest: The authors declare no conflict of interest.

References

1. Ordoudi, S.A.; Tsermentseli, S.K.; Nenadis, N.; Assimopoulou, A.N.; Tsimidou, M.Z.; Papageorgiou, V.P. Structure-radical scavenging activity relationship of alkannin/shikonin derivatives. *Food Chem.* **2011**, *124*, 171–176. [CrossRef]
2. Papageorgiou, V.P. Naturally occurring isohexenylnaphthazarin pigments: A new class of drugs. *Planta Med.* **1980**, *38*, 193–203. [CrossRef] [PubMed]
3. Papageorgiou, V.P.; Assimopoulou, A.N.; Ballis, A.C. Alkannins and shikonins: A new class of wound healing agents. *Curr. Med. Chem.* **2008**, *15*, 3248–3267. [CrossRef] [PubMed]
4. Papageorgiou, V.P.; Assimopoulou, A.N.; Couladouros, E.A.; Hepworth, D.; Nicolaou, K.C. The chemistry and biology of alkannin, shikonin, and related naphthazarin natural products. *Angew. Chem. Int. Ed.* **1999**, *38*, 270–300. [CrossRef]
5. Papageorgiou, V.P.; Assimopoulou, A.N.; Samanidou, V.F.; Papadoyannis, I.N. Recent advances in chemistry, biology and biotechnology of alkannins and shikonins. *Curr. Org. Chem.* **2006**, *10*, 2123–2142. [CrossRef]
6. Karapanagioti, E.G.; Assimopoulou, A.N. Naturally occurring wound healing agents: An evidence-based review. *Curr. Med. Chem.* **2016**, *23*, 3285–3321. [CrossRef] [PubMed]
7. Xu, Z.; Jia-Hua, C.; Qing-Qing, M.; Shao-Shun, L.; Wen, Z.; Sui, X. Advance in anti-tumor mechanisms of shikonin, alkannin and their derivatives. *Mini-Rev. Med. Chem.* **2018**, *18*, 164–172.
8. Xie, Y.; Hou, X.L.; Wu, C.L. The research progress of cell apoptosis induced by shikonin and signal pathway of apoptosis. *Chin. J. Pharm. Biotechnol.* **2016**, *23*, 173–178.
9. Zhao, Q.; Assimopoulou, A.N.; Klauck, S.M.; Damianakos, H.; Chinou, I.; Kretschmer, N.; Rios, J.L.; Papageorgiou, V.P.; Bauer, R.; Efferth, T. Inhibition of c-MYC with involvement of ERK/JNK/MAPK and AKT pathways as a novel mechanism for shikonin and its derivatives in killing leukemia cells. *Oncotarget* **2015**, *6*, 38934–38951. [CrossRef] [PubMed]
10. Zhang, F.-Y.; Hu, Y.; Que, Z.-Y.; Wang, P.; Liu, Y.-H.; Wang, Z.-H.; Xue, Y.-X. Shikonin inhibits the migration and invasion of human glioblastoma cells by targeting phosphorylated β-catenin and phosphorylated PI3K/Akt: A potential mechanism for the anti-glioma efficacy of a traditional chinese herbal medicine. *Int. J. Mol. Sci.* **2015**, *16*, 23823–28848. [CrossRef] [PubMed]

11. Duan, D.; Zhang, B.; Yao, J.; Liu, Y.; Fang, J. Shikonin targets cytosolic thioredoxin reductase to induce ROS-mediated apoptosis in human promyelocytic leukemia HL-60 cells. *Free Radic. Biol. Med.* **2014**, *70*, 182–193. [CrossRef] [PubMed]
12. Wiench, B.; Eichhorn, T.; Paulsen, M.; Efferth, T. Shikonin directly targets mitochondria and causes mitochondrial dysfunction in cancer cells. *Evid.-Based Complement. Altern. Med.* **2012**, *2012*, 726025. [CrossRef] [PubMed]
13. Spyrelli, E.D.; Kyriazou, A.V.; Virgiliou, C.; Nakas, A.; Deda, O.; Papageorgiou, V.P.; Assimopoulou, A.N.; Gika, H.G. Metabolic profiling study of shikonin's cytotoxic activity in the Huh7 human hepatoma cell line. *Mol. BioSyst.* **2017**, *13*, 841–851. [CrossRef] [PubMed]
14. Zhou, Q. *Shikonin and NRF2 Chemoprevention*; University of Maryland Baltimore; National Cancer Institute: Baltimore, MD, USA, 2014. Available online: http://projectreporter.nih.gov/project_info_description.cfm?aid=8685189 (accessed on 25 May 2018).
15. Ni, F.; Huang, X.; Chen, Z.; Qian, W.; Tong, X. Shikonin exerts antitumor activity in Burkitt's lymphoma by inhibiting C-MYC and PI3K/AKT/mTOR pathway and acts synergistically with doxorubicin. *Sci. Rep.* **2018**, *8*, 3317. [CrossRef] [PubMed]
16. Kozako, T.; Arima, N.; Yoshimitsu, M.; Honda, S.I.; Soeda, S. Liposomes and nanotechnology in drug development: Focus on oncotargets. *Int. J. Nanomed.* **2012**, *7*, 4943–4951. [CrossRef] [PubMed]
17. Alexis, F.; Rhee, J.W.; Richie, J.P.; Radovic-Moreno, A.F.; Langer, R.; Farokhzad, O.C. New frontiers in nanotechnology for cancer treatment. *Urol. Oncol. Semin. Orig. Investig.* **2008**, *26*, 74–85. [CrossRef] [PubMed]
18. Gregoriadis, G.; Wills, E.J.; Swain, C.P.; Tavill, A.S. Drug-carrier potential of liposomes in cancer chemotherapy. *Lancet* **1974**, *1*, 1313–1316. [CrossRef]
19. Gumulec, J.; Fojtu, M.; Raudenska, M.; Sztalmachova, M.; Skotakova, A.; Vlachova, J.; Skalickova, S.; Nejdl, L.; Kopel, P.; Knopfova, L.; et al. Modulation of induced cytotoxicity of doxorubicin by using apoferritin and liposomal cages. *Int. J. Mol. Sci.* **2014**, *15*, 22960–22977. [CrossRef] [PubMed]
20. Heger, Z.; Polanska, H.; Merlos Rodrigo, M.A.; Guran, R.; Kulich, P.; Kopel, P.; Masarik, M.; Eckschlager, T.; Stiborova, M.; Kizek, R.; et al. Prostate tumor attenuation in the nu/nu murine model due to anti-sarcosine antibodies in folate-targeted liposomes. *Sci. Rep.* **2016**, *6*, 33379. [CrossRef] [PubMed]
21. Li, J.; Wang, X.; Zhang, T.; Wang, C.; Huang, Z.; Luo, X.; Deng, Y. A review on phospholipids and their main applications in drug delivery systems. *Asian J. Pharm. Sci.* **2015**, *10*, 81–98. [CrossRef]
22. Bulbake, U.; Doppalapudi, S.; Kommineni, N.; Khan, W. Liposomal formulations in clinical use: An updated review. *Pharmaceutics* **2017**, *9*, 12. [CrossRef] [PubMed]
23. Woodle, M.C.; Lasic, D.D. Sterically stabilized liposomes. *Biochim. Biophys. Acta* **1992**, *1113*, 171–199. [CrossRef]
24. Drummond, D.C.; Meyer, O.; Hong, K.; Kirpotin, D.B.; Papahadjopoulos, D. Optimizing liposomes for delivery of chemotherapeutic agents to solid tumors. *Pharmacol. Rev.* **1999**, *51*, 691–744. [PubMed]
25. Andresen, T.L.; Jensen, S.S.; Jørgensen, K. Advanced strategies in liposomal cancer therapy: Problems and prospects of active and tumor specific drug release. *Prog. Lipid Res.* **2005**, *44*, 68–97. [CrossRef] [PubMed]
26. Petersen, G.H.; Alzghari, S.K.; Chee, W.; Sankari, S.S.; La-Beck, N.M. Meta-analysis of clinical and preclinical studies comparing the anticancer efficacy of liposomal versus conventional non-liposomal doxorubicin. *J. Control. Release* **2016**, *232*, 255–264. [CrossRef] [PubMed]
27. Xia, H.; Tang, C.; Gui, H.; Wang, X.; Qi, J.; Wang, X.; Yang, Y. Preparation, cellular uptake and angiogenic suppression of shikonin-containing liposomes in vitro and in vivo. *Biosci. Rep.* **2013**, *33*, 207–215. [CrossRef] [PubMed]
28. Kontogiannopoulos, K.N.; Assimopoulou, A.N.; Dimas, K.; Papageorgiou, V.P. Shikonin–loaded liposomes as a new drug delivery system: Physicochemical characterization and in vitro cytotoxicity. *Eur. J. Lipid Sci. Technol.* **2011**, *113*, 1113–1123. [CrossRef]
29. Kontogiannopoulos, K.N.; Tsermentseli, S.K.; Assimopoulou, A.N.; Papageorgiou, V.P. Sterically stabilized liposomes as a potent carrier for shikonin. *J. Liposome Res.* **2014**, *24*, 230–240. [CrossRef] [PubMed]
30. Han, H.K.; Shin, H.J.; Ha, D.H. Improved oral bioavailability of alendronate via the mucoadhesive liposomal delivery system. *Eur. J. Pharm. Sci.* **2012**, *46*, 500–507. [CrossRef] [PubMed]
31. Crosasso, P.; Ceruti, M.; Brusa, P.; Arpicco, S.; Dosio, F.; Cattel, L. Preparation, characterization and properties of sterically stabilized paclitaxel-containing liposomes. *J. Control. Release* **2000**, *63*, 19–30. [CrossRef]

32. Kelly, C.; Jefferies, C.; Cryan, S.A. Targeted liposomal drug delivery to monocytes andmacrophages. *J. Drug Deliv.* **2011**, *2011*, 727241. [CrossRef] [PubMed]
33. Assimopoulou, A.N.; Ganzera, M.; Stuppner, H.; Papageorgiou, V.P. Simultaneous determination of monomeric and oligomeric alkannins and shikonins by high-performance liquid chromatography–diode array detection–mass spectrometry. *Biomed. Chromatogr.* **2008**, *22*, 173–190. [CrossRef] [PubMed]
34. Krasnici, S.; Werner, A.; Eichhorn, M.E.; Schmitt-Sody, M.; Pahernik, S.A.; Sauer, B.; Schulze, B.; Teifel, M.; Michaelis, U.; Naujoks, K.; et al. Effect of the surface charge of liposomes on their uptake by angiogenic tumor vessels. *Int. J. Cancer* **2003**, *105*, 561–567. [CrossRef] [PubMed]
35. Cho, E.C.; Lim, H.J.; Shim, J.; Kim, J.; Chang, I.S. Improved stability of liposome in oil/water emulsion by association of amphiphilic polymer with liposome and its effect on bioactive skin permeation. *Colloids Surf. A Physicochem. Eng. Asp.* **2007**, *299*, 160–168. [CrossRef]
36. Gardikis, K.; Hatziantoniou, S.; Bucos, M.A.; Fessas, D.; Signorelli, M.; Felekis, T.; Zervou, M.; Screttas, C.G.; Steele, B.R.; Ionov, M.; et al. New drug delivery nanosystem combining liposomal and dendrimeric technology (liposomal locked-in dendrimers) for cancer therapy. *J. Pharm. Sci.* **2010**, *99*, 3561–3571. [CrossRef] [PubMed]
37. Bonacucina, G.; Cespi, M.; Misici-Falzi, M.; Palmieri, G.F. Colloidal soft matter as drug delivery system. *J. Pharm. Sci.* **2009**, *98*, 1–42. [CrossRef] [PubMed]
38. Zhou, L. *Guidance for Industry Liposome Drug Products. Chemistry, Manufacturing, and Controls; Human Pharmacokinetics and Bioavailability; and Labeling Documentation*; Center for Drug Evaluation and Research (CDER), U.S. Department of Health and Human Services, Food and Drug Administration: Washington, DC, USA, 2002.
39. Tenzel, R.A.; Aitcheson, D.F. Preparation of uniform-size liposomes and other lipid structures. WO1989011335A1, 30 November 1989.
40. Dadashzadeh, S.; Mirahmadi, N.; Babaei, M.H.; Vali, A.M. Peritoneal retention of liposomes: Effects of lipid composition, PEG coating and liposome charge. *J. Control. Release* **2010**, *148*, 177–186. [CrossRef] [PubMed]
41. Sriwongsitanont, S.; Ueno, M. Effect of a PEG lipid (DSPE-PEG2000) and freeze-thawing process on phospholipid vesicle size and lamellarity. *Colloid Polym. Sci.* **2004**, *282*, 753–760. [CrossRef]
42. Shenoy, V.S.; Gude, R.P.; Murthy, R.S.R. Investigations on paclitaxel loaded HSPC based conventional and PEGylated liposomes: In vitro release and cytotoxic studies. *Asian J. Pharm. Sci.* **2011**, *6*, 1–7.
43. Muller, R.H. *Colloidal Carriers for Controlled Drug Delivery and Targeting: Modification, Characterization, and in vivo Distribution*; CRC Press: Boca Raton, FL, USA, 1991.
44. Gentile, E.; Cilurzo, F.; Di Marzio, L.; Carafa, M.; Ventura, C.A.; Wolfram, J.; Paolino, D.; Celia, C. Liposomal chemotherapeutics. *Future Oncol.* **2013**, *9*, 1849–1859. [CrossRef] [PubMed]
45. Mohammed, A.R.; Weston, N.; Coombes, A.G.A.; Fitzgerald, M.; Perrie, Y. Liposome formulation of poorly water soluble drugs: Optimisation of drug loading and ESEM analysis of stability. *Int. J. Pharm.* **2004**, *285*, 23–34. [CrossRef] [PubMed]
46. Yang, T.; Cui, F.D.; Choi, M.K.; Cho, J.W.; Chung, S.J.; Shim, C.K.; Kim, D.D. Enhanced solubility and stability of PEGylated liposomal paclitaxel: In vitro and in vivo evaluation. *Int. J. Pharm.* **2007**, *338*, 317–326. [CrossRef] [PubMed]
47. Vali, A.M.; Toliyat, T.; Shafaghi, B.; Dadashzadeh, S. Preparation, optimization, and characterization of topotecan loaded PEGylated liposomes using factorial design. *Drug Dev. Ind. Pharm.* **2008**, *34*, 10–23. [CrossRef] [PubMed]
48. Mourelatou, E.A.; Libster, D.; Nir, I.; Hatziantoniou, S.; Aserin, A.; Garti, N.; Demetzos, C. Type and location of interaction between hyperbranched polymers and liposomes. Relevance to design of a potentially advanced drug delivery nanosystem (aDDnS). *J. Phys. Chem. B* **2011**, *115*, 3400–3408. [CrossRef] [PubMed]
49. Fatouros, D.G.; Antimisiaris, S.G. Effect of amphiphilic drugs on the stability and zeta-potential of their liposome formulations: A study with prednisolone, diazepam, and griseofulvin. *J. Colloid Interface Sci.* **2002**, *251*, 271–277. [CrossRef] [PubMed]
50. Dadashzadeh, S.; Vali, A.M.; Rezaie, M. The effect of PEG coating on in vitro cytotoxicity and in vivo disposition of topotecan loaded liposomes in rats. *Int. J. Pharm.* **2008**, *353*, 251–259. [CrossRef] [PubMed]
51. Wang, C.H.; Huang, Y.Y. Encapsulating protein into preformed liposomes by ethanol-destabilized method. *Artif. Cells Nanomed. Biotechnol.* **2003**, *31*, 303–312. [CrossRef]

52. Abra, R.M.; Mihalko, P.J.; Schreier, H. The effect of lipid composition upon the encapsulation and in vitro leakage of metaproterenol sulfate from 0.2 µm diameter, extruded, multilamellar liposomes. *J. Control. Release* **1990**, *14*, 71–78. [CrossRef]
53. Colletier, J.P.; Chaize, B.; Winterhalter, M.; Fournier, D. Protein encapsulation in liposomes: Efficiency depends on interactions between protein and phospholipid bilayer. *BMC Biotechnol.* **2002**, *2*, 9. [CrossRef]
54. Haeri, A.; Alinaghian, B.; Daeihamed, M.; Dadashzadeh, S. Preparation and characterization of stable nanoliposomal formulation of fluoxetine as a potential adjuvant therapy for drug-resistant tumors. *Iran. J. Pharm. Res.* **2014**, *13*, 3–14. [PubMed]
55. Ho, E.A.; Osooly, M.; Strutt, D.; Masin, D.; Yang, Y.; Yan, H.; Bally, M. Characterization of long-circulating cationic nanoparticle formulations consisting of a two-stage PEGylation step for the delivery of siRNA in a breast cancer tumor model. *J. Pharm. Sci.* **2013**, *102*, 227–236. [CrossRef] [PubMed]
56. Tan, S.; Li, X.; Guo, Y.; Zhang, Z. Lipid-enveloped hybrid nanoparticles for drug delivery. *Nanoscale* **2013**, *5*, 860–872. [CrossRef] [PubMed]
57. Begum, M.Y.; Abbulu, K.; Sudhakar, M. Preparation, characterization and in-vitro release study of flurbiprofen loaded stealth liposomes. *Chem. Sci. Trans.* **2012**, *1*, 201–209. [CrossRef]
58. Pippa, N.; Psarommati, F.; Pispas, S.; Demetzos, C. The shape/morphology balance: A study of stealth liposomes via fractal analysis and drug encapsulation. *Pharm. Res.* **2013**, *30*, 2385–2395. [CrossRef] [PubMed]

fluids

MDPI

Article

Pipette Petri Dish Single-Cell Trapping (PP-SCT) in Microfluidic Platforms: A Passive Hydrodynamic Technique

Vigneswaran Narayanamurthy [1,2,*], Tze Pin Lee [3], Al'aina Yuhainis Firus Khan [4], Fahmi Samsuri [1], Khairudin Mohamed [3], Hairul Aini Hamzah [2] and Madia Baizura Baharom [2]

1 Faculty of Electrical and Electronics Engineering, University Malaysia Pahang Pekan, Pekan, Pahang 26600, Malaysia; fahmi@ump.edu.my

2 Kulliyyah of Medicine, International Islamic University Malaysia, Kuantan, Pahang 25200, Malaysia; hairulaini@iium.edu.my (H.A.H.); madia@iium.edu.my (M.B.B.)

3 Nanofabrication and Functional Materials Research Group, School of Mechanical Engineering, University Science Malaysia, Engineering Campus, Nibong Tebal, Penang 14300, Malaysia; tpin90@hotmail.com (T.P.L.); mekhairudin@usm.my (K.M.)

4 Kulliyyah of Allied Health Science, International Islamic University Malaysia, Kuantan, Pahang 25200, Malaysia; alainayuhainis@gmail.com

* Correspondence: PEL13006@stdmail.ump.edu.my; Tel.: +6-010-536-9493

Received: 20 June 2018; Accepted: 22 July 2018; Published: 24 July 2018

Abstract: Microfluidics-based biochips play a vital role in single-cell research applications. Handling and positioning of single cells at the microscale level are an essential need for various applications, including genomics, proteomics, secretomics, and lysis-analysis. In this article, the pipette Petri dish single-cell trapping (PP-SCT) technique is demonstrated. PP-SCT is a simple and cost-effective technique with ease of implementation for single cell analysis applications. In this paper a wide operation at different fluid flow rates of the novel PP-SCT technique is demonstrated. The effects of the microfluidic channel shape (straight, branched, and serpent) on the efficiency of single-cell trapping are studied. This article exhibited passive microfluidic-based biochips capable of vertical cell trapping with the hexagonally-positioned array of microwells. Microwells were 35 μm in diameter, a size sufficient to allow the attachment of captured cells for short-term study. Single-cell capture (SCC) capabilities of the microfluidic-biochips were found to be improving from the straight channel, branched channel, and serpent channel, accordingly. Multiple cell capture (MCC) was on the order of decreasing from the straight channel, branch channel, and serpent channel. Among the three designs investigated, the serpent channel biochip offers high SCC percentage with reduced MCC and NC (no capture) percentage. SCC was around 52%, 42%, and 35% for the serpent, branched, and straight channel biochips, respectively, for the tilt angle, θ values were between 10–15°. Human lung cancer cells (A549) were used for characterization. Using the PP-SCT technique, flow rate variations can be precisely achieved with a flow velocity range of 0.25–4 m/s (fluid channel of 2 mm width and 100 μm height). The upper dish (UD) can be used for low flow rate applications and the lower dish (LD) for high flow rate applications. Passive single-cell analysis applications will be facilitated using this method.

Keywords: hydrodynamics; microfluidics; pipette Petri dish single-cell trapping (PP-SCT); passive trapping; single-cell trapping; single cell analysis; tilt trapping

1. Introduction

Cytology has been extensively studied since the invention of the microscope. However, in recent years, microfluidics has predominantly come into play for single-cell analysis (SCA) due to the development of modern fabrication technologies and tools [1–3]. The increase of sensitivity of many analytical systems in cell and fluid handling are attributed to the microfluidics technology, which has pushed the SCA to an advanced level. The number of articles published on SCA has increased manifold as compared to past decades. The main reason for the ascendancy of using SCA is because it will reduce the biological errors from the target cell population [4]. SCA acts as a tool to clarify molecular mechanisms and pathways revealing the nature of cell heterogeneity. Moreover, each cell exhibits different cell cycle stages, protein, and gene expressions [5]. Apart from that, SCA facilitates rare cells or events, scarce, precious samples, and single-cell precision in populations.

1.1. Microfluidics

Single cells are to be trapped and isolated to perform single-cell investigations [6–10]. Microfluidics has come out as a powerful tool to study the complexities of cells [11]. Apart from the quality of results, the reduction in time, savings in space, reagents and small sample volumes involved (on the order of microliters) are some of the remarkable discoveries made using microfluidics technology [4]. Several approaches [12–22] were utilized in microfluidics facilitating single-cell trapping which can be broadly classified as contact-based and contactless. Various cell trapping methods have been developed using different approaches including chemical, gel, hydrodynamic, dielectrophoresis, magnetic, laser, and acoustics [23–26]. Every method has its merits and demerits. Usually, the process requires auxiliaries, like a pump or a pressure-controlling system for fluid introduction and guidance. Some processes need specialized electronics or optical equipment. This specialized auxiliary equipment is not commonly used in medical or biological laboratories, therefore, inducing additional access-related obstacles for operation. The area density of single-cell arrays is reduced significantly due to the space occupation by the auxiliary parts or the devices. Usually, the design structures of the microfluidic devices require the fabrication of many parts including valves, multiple layers, and channels, which are complicated [27–29]. Thereby the practical applications in clinics and general biological laboratories are limited due to these downsides [24]. These downsides have led to the need for the development of a hydrodynamic trapping technique where the trapping is performed using the design itself by involving unsophisticated basic laboratory instruments for single cell measurements. The hydrodynamic trapping technique is also capable of preserving and handling the cells which may be affected due to the presence of an external field/force. Apart from the simplicity of the process, the natural state of the separated particles can be of use for several other further studies.

1.2. Hydrodynamic Trapping

In hydrodynamic trapping approach, the target particles from the main flow are separated using the mechanical obstacles or barriers. Once separated, the target particles are retained in hydrodynamic trapping sites where various investigations can be performed. Walls and pores are the typically-used particle trapping structures, arrays of these trapping pattern structures can be fabricated to capture the single cells. SCA using labeling techniques can improve significantly the specificity and sensitivity. The most common label is a fluorescent molecule, which is used to visualize the analyte. Therefore, the combination of microfluidic platforms and multimodal spectroscopy opens the gate for next-generation single-cell studies. With that combo, SCA with defined stimulation of cells and analysis of cell responses can be performed simultaneously for as many different bio-molecules as possible. This will also address several current needs and demands for technologies identified in the literature to pursue SCA [5,30]. Thus, the field is a matter of interest for analytical and diagnostic aspects.

Several researchers are working on hydrodynamic-based single-cell trapping [24,26,31–35]. New smart methods and platforms facilitating multiple measurements on individual cells are entailed. Development of a platform, concurrently having high throughput and resolution with less expertise requirement remains a challenge. The intent here is to facilitate a less complex microfluidic device, which helps in vertical cell trapping based on the design and technique itself. Additionally, the process involving unsophisticated basic laboratory instruments for single-cell measurements, which are compatible with conventional fluorescence imaging modalities are the current needs of SCA platforms. In this article a simple, but efficient, pipette Petri dish single-cell trapping technique (PP-SCT) is demonstrated. This topic can also be termed as tilt trapping (TT). The PP-SCT technique only involves basic lab equipment—as the name suggests, a pipette and a Petri dish—thereby its implementation and application scope is very wide. Thus, this method can be easily implemented for single-cell analysis without the need of complex auxiliaries.

In 2005, Rettig and Folch demonstrated the optimization of high single-cell trapping in microwells by deriving a relation between well depth, well diameter, and settling time [12]. In 2010, Park et al. demonstrated the microwell design analysis for efficient single-cell capture (SCC) [14]. In 2013, Karimi et al. reported an overview of the cell/particle sorting techniques using hydrodynamic effects in microchannels [36]. To the best of our knowledge, none of the studies in the literature have investigated the relative efficiency of hydrodynamic-based fluid channel designs on SCC. In this article, the effects of fluid channel design and microwell array design orientation for single-cell trapping efficiency using the PP-SCT technique has been investigated. Three designs were considered straight channel, branched channel, and serpent shaped channel. A simple, but efficient, microfluid dynamics-based passive technique, capable of vertical cell trapping with high SCC for short-term SCA study was demonstrated.

2. Methods

2.1. Concept and Design Criteria

Figure 1 shows the illustration of the PP-SCT technique. In this technique, the inlet edge of the biochip is placed over the side wall of the Petri dish, and the outlet edge is placed inside the Petri dish base along its diameter. When the biochip is moved externally to the Petri dish, the angle and height of the inlet are varied. In this research, a glass slide biochip 3 inches × 1 inches in size, and standard Petri dishes of 90 mm × 15 mm (Sigma, St Louis, MO, USA) were considered. Petri dishes have a lower dish (LD) around 87 mm × 15 mm and an upper dish (UD) around 90 mm × 7 mm. The cell suspension was pipetted out into the inlet well, which flows through the channel to be collected at the outlet well. During the process, the cell gets captured in the microwell array. Biochip designs with the straight channel, branched channel and serpent fluid channel were designed, fabricated and investigated. Figure 2 shows the shape of different fluid channels considered. Sticky tape can be used to lock the biochip position with the Petri dish during the process, if operated at extreme ranges.

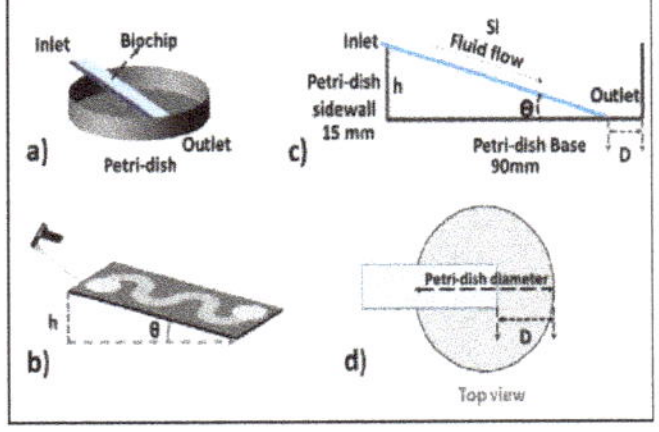

Figure 1. (**a**,**b**) Illustration of pipette Petri dish single-cell trapping (PP-SCT); (**c**,**d**) shows the side and top view, respectively, where θ is the tilt angle, h is the height of the inlet, D is the distance between biochip outlet side edge to the Petri dish wall along the diameter, and Sl is glass slide length.

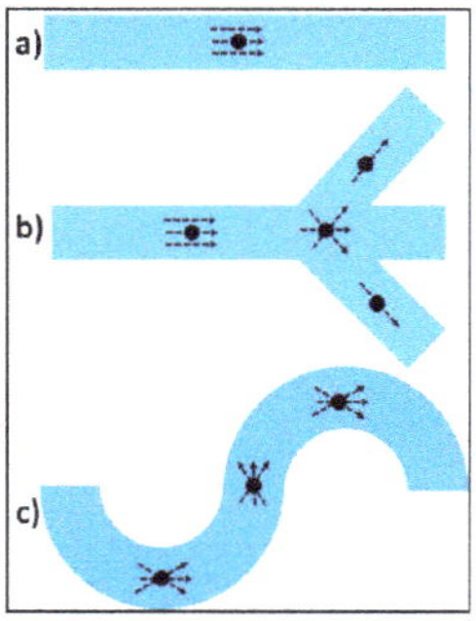

Figure 2. Fluid channel shapes (**a**) straight channel; (**b**) branched channel, and (**c**) serpent-shaped channel.

Considering a fluid flow from an elevated height to a lower height, by law of conservation of energy, fluid velocity (V) at the outlet neglecting the frictional losses can be given as:

$$V = \sqrt{(2g(h1 - h2))} \quad (1)$$

In the above equation, g is the gravitational constant. Position 1 and position 2 corresponds to the inlet and outlet, and h1 and h2 correspond to the inlet and outlet elevated height, respectively. Flow rate (Q) can be given as:

$$Q = A \times V \quad (2)$$

where A is the cross-sectional area of the fluid channel and V is the fluid velocity. Both the fluid velocity and flow rate are adjustable through alteration of the elevation head, which can be simply done by altering the D in the PP-SCT method. $\Theta \propto D \propto h$ (for θ values between 0° and 90°). Figure 3 shows the varying operational modes of the PP-SCT technique and its characteristics.

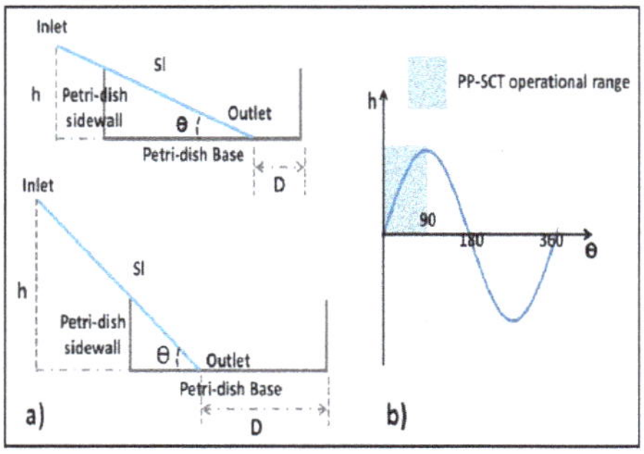

Figure 3. (**a**) Shows the varying operational position using the PP-SCT technique; (**b**) shows the relation between the angles of elevation to the inlet elevations (height), and height follows the sine wave with respect to the angle.

2.2. Computational Analysis

PP-SCT was modeled and simulated using Comsol Multiphysics solver (version 5.0, COMSOL AB., Burlington, MA, USA). The simulation was carried out using water as the flow material. Newtonian and single-phase fluid models were considered, by taking into account the law of conservation of energy, Stoke's law, and the continuity equation [37]. The Navier-Stokes equations govern the fluid motion and can be seen as Newton's second law of motion for fluids:

$$\rho((\partial u/\partial t) + u\cdot\nabla u) = -\nabla p + \nabla\cdot(\mu(\nabla u + (\nabla T\, u)) - 2/3(\mu(\nabla\cdot u)\, I)) + F \quad (3)$$

where u is the fluid velocity, ρ is the fluid density, p is the fluid pressure, and μ is the fluid dynamic viscosity. The equation includes the inertial forces, pressure forces, viscous forces, and the external forces applied to the fluid. These equations are always solved together with the continuity equation:

$$\partial\rho/\partial t + \nabla\cdot(\rho u) = 0 \quad (4)$$

The Navier-Stokes equations represent the conservation of momentum, while the continuity equation represents the conservation of mass. For incompressible flows, the continuity equation yields:

$$\nabla\cdot u = 0 \quad (5)$$

$$0 = -\nabla_t\, p\cdot e_t - 1/2((f_d\cdot\rho/d_h)\,|\,u\,|\,u) + F\cdot e_t \quad (6)$$

$$-\nabla_t\cdot(A\cdot\rho\cdot\mu\cdot e_t) = 0 \quad (7)$$

where e_t is the tangential unit vector along the edge.

The flow was modeled for a rectangular straight channel with a width of 2 mm and a height of 100 μm, for different angles, and its corresponding fluid velocity was plotted. During the simulation, the Darcy friction factor for the rectangular microchannel with glass surface roughness of 0.0015 mm was given as a boundary condition. A grid-independent study was carried out and the results suggest that the finer and coarser mesh can produce grid-independent solutions. As a result, the computation results are obtained using the physics-controlled grid system. Table 1 lists the properties and constants used for the simulation. The Darcy friction factor f_d using the Darcy–Weisbach equation is given below:

$$f_d = 64/\mathrm{Re} \quad (8)$$

The Reynolds number [38,39] is given by the equation:

$$\mathrm{Re} = (LV_{avg}\cdot\rho)/\mu \quad (9)$$

where V_{avg} is the average fluid velocity, and L is the relevant fluid length scale, for rectangular channel:

$$L = 4A/P \quad (10)$$

where A and P are the cross-section area and wetted perimeter of the channel, respectively. Along with that, the gravitational volumetric force was considered.

Table 1. Properties, constants, and boundary conditions used for the simulation.

Components	Property	Value/Equation
Fluid (water)	Dynamic Viscosity (mu) Density (rho)	8.90×10^{-4} (Pa·s) 997 (kg/m^3)
Channel	Channel Surface Roughness	Glass (0.0015 mm)
Initial Values	Pressure Tangential Velocity	101,325 (Pa) 0 m/s
Volume Force	x y	0 (N/m^3) (−g_const) * pfl.rho (N/m^3)
Pressure	Inlet Outlet	101,325 (Pa) 101,325 (Pa) + (5(mm)) * g_const * pfl.rho

pfl is the fluid flow physics model in Comsol, g_const is the gravitational constant, * indicates the multiplication and . indicates model based properties used in Comsol.

2.3. Biochip Fabrication

Biochips were fabricated using the emulsion mask grayscale photolithography process [40]. The biochip patterns were designed using CorelDraw X7. The designed pattern was drawn five times larger than the original dimension because of the 5:1 shrinking rate of the mask fabrication equipment. The percentage of grayscale concentrations were used to control the height of the developed photoresist obtained after the ultraviolet (UV) exposure process. The designed grayscale mask was printed on a transparent polyethylene terephthalate (PET) film by using the image setter technique. The printed transparent film can also be called a master mask film. To project the image from the printed master mask film in a 5:1 reduction scale onto a high-precision photo plate (Konica Minolta, Inc., Tokyo, Japan) also known as an emulsion mask, an MM605 simple mask fabrication machine (Nanometric Technology Inc., Milpitas, CA, USA) was used. The light-sensitive silver halide was coated over the emulsion mask. Due to the high light sensitivity of the emulsion mask, the whole mask exposure process was performed under dark room conditions. Before the emulsion mask exposure process, the silver halide-coated surface was placed facing towards the light source with the exposure time adjusted to 8 s. The exposed emulsion mask was immersed into an emulsion mask developer at room temperature for 2 min after the exposure process. A mixture of one part of high-resolution plate developer (CDH-100) from Konica Minolta Opto, Inc., Osaka, Japan and four parts of distilled water was used as the emulsion mask developer. To ensure the uniform development process, the immersed emulsion mask was stirred continuously. The previously-exposed silver halide had formed a high optical density metallic silver during the development process, which can function as an excellent optical filter on the emulsion mask. During the development process, the darkfield emulsion mask was created.

The 3 × 1 inch-sized microscope glass slides (DURAN Group, DWK Life Sciences GmbH, Mainz, Germany) were used as a substrate. The glass substrate was chosen due to its transparent property to produce a 3D relief surface structure by employing the back UV exposure process. The glass slides were washed to remove surface contaminants by using an ultrasonic bath (GT Sonic VGT-1613QTD, GuangDong GT Ultrasonic Co. Ltd., Guangdong, China). First, the washing process was done by an ultrasonic bath with acetone for 5 min, followed by an ultrasonic bath with methanol and isopropyl alcohol (IPA) each for 5 min. After the above washing process, the glass slides were rinsed with distilled water. Quickly after the surface washing process, 2 mL of SU-8 2010 photoresist (MicroChem, Newton, MA, USA) was dispensed and spread over the whole surface of the 3 × 1-inch glass substrate to generate a 700-μm thick SU-8 film. During this process, the coated SU-8 film was self-planarized by itself. The obtained resultants were a flat and uniform layer due to the surface tension and high mobility. The SU-8-coated glass substrate was soft baked using a conventional oven at a temperature of 95 °C for 10 h. It is essential to place all the SU-8-coated samples flat in the conventional oven to avoid any gravity force affecting the flow of the photoresist. Then, the LA4100_R1 one-side mask aligner (Sanei Electric Inc., Tokyo, Japan) was used to back expose the photoresist-coated glass substrates. Figure 4 illustrates the grayscale photolithography process. The SU-8-coated glass substrates were exposed using a 180 W mercury lamp of 365 nm for 30 s. A post-exposure bake was conducted in three steps immediately following the UV exposure process. Firstly, the exposed samples were baked at a temperature of 65 °C for 2 min on a hotplate. Secondly, the temperature of the hotplate was gradually ramped up to 95 °C for 10 min. Lastly, after 10 min of baking the hotplate was switched off while the samples were still left on top of the hotplate. The samples were then allowed to cool down gently to room temperature. This slow cooling process decreases the stress built up in the cross-linked SU-8 and, thereby, avoids cracking and deformation of patterns during the development process. The developing process was performed by immersing the exposed samples into SU-8 developer solution (MicroChem, Newton, MA, USA). The development process was enhanced by using an ultrasonic cleaner and took only 2 min to dissolve all the unexposed SU-8 photoresist from the glass substrate completely. After the first development, the developed samples were washed by using another batch of clean SU-8 developer solution, then followed by the IPA and distilled water. Finally, a stream of nitrogen gas was used to blow dry the developed samples. Fabricated biochips are shown in Figure 5. The biochip consists of a fluid channel with a width of 2 mm, and with a hexagonally positioned microwell array with a diameter around 35 μm and a depth of 30 μm. The straight channel design consists of ~3500 microwells, the branched channel design consists of ~5000 microwells, and the serpent shaped design consists of ~1100 microwells. There was a −5% change in the feature size of the fabricated biochips compared to the original design.

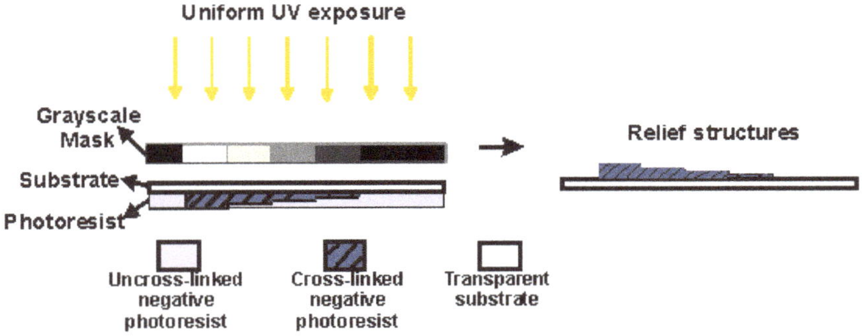

Figure 4. Illustration of the grayscale photolithography process.

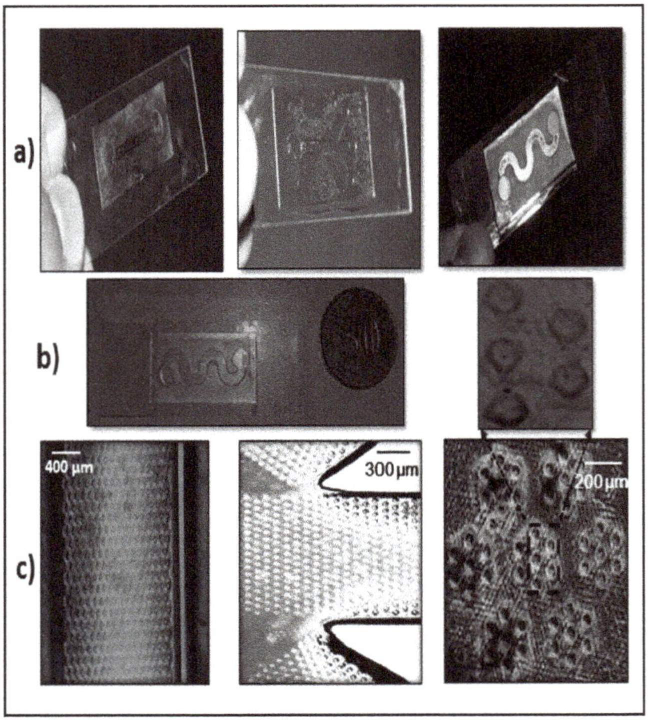

Figure 5. (**a**) Fabricated biochips; (**b**) biochip scale with a Malaysian 50 sen coin used as a reference; and (**c**) a microwell array on the fabricated biochip, images taken from an infinite focus measurement system (ALICONA, Graz, Austria).

2.4. Cell Culture

Human A549 lung cancer cells were obtained from ATCC (Rockville, MD, USA). Cells were cultured and maintained in complete growth media (CGM) consisting of 89% Dulbecco's Modified Eagle's Medium (Gibco, Grand Island, NY, USA), 10% fetal bovine serum (Gibco, Grand Island, NY, USA), and 1% of penicillin-streptomycin antibiotic (Gibco, Grand Island, NY, USA). The cells were maintained at 37 °C under a humidified atmosphere, with 5% CO_2 and 95% air, respectively.

2.5. Single-Cell Trapping and Cell Viability Tests

Prior to operation, the biochip was covered with a coverslip and sealed with sticky tape. Biochips were sterilized with 70% ethanol and UV for a few minutes before using. Once the grown cells attained 80% confluency, they were trypsinized with TrypLe (Gibco, Grand Island, NY, USA) and centrifuged (Eppendorf 5810, Westbury, NY, USA) at 700 rpm for 5 min. The cell pellet was resuspended in complete growth media. Cell suspension with a density of 3.5×10^4 to 7.5×10^4 cells/mL were mixed with Trypan Blue (Sigma-Aldrich, St Louis, MO, USA) and were pipetted down into the biochip, which was tilted at an angle of around 10–15° (D was around 16–30 mm for LD). By varying the D along with the ruler, the angle was varied accordingly. Pipetting was done gently and slowly. After the cell-suspended growth media was partially drained out from the outlet and the channels were filled with cells and growth media, the biochip was gently placed into a Petri dish and incubated. During the transfer to the Petri dish, caution was taken not to shake the biochip, as it may result in moving the cell out of the microwell before attachment. Additional growth media can be added if required after placing into Petri dish. The microwell array was observed under a microscope (Leica DM 2000 LED, Bensheim, Germany) for its trapping and viability. The overall technique was kept simple, and not much complex process handling was done on the cells. Statistical analysis was performed using GraphPad Prism 6.0. ANNOVA analysis was carried out to find the significant factors within each channel design and two-way ANNOVA analysis for comparing between channel designs. * is $p < 0.01$, ** is $p < 0.001$, *** $p < 0.0001$, and **** is $p < 0.00001$, indicating that the result is significant.

2.6. Single-Cell Fluorescent Measurement

After loading the cells, the biochip was incubated at 37 °C under a humidified atmosphere, with 5% CO_2 and 95% air for 3 h for cell attachment. Once attached, the biochip was washed with the phosphate-buffered solution (PBS) to fully remove remaining media. Cells were fixed with 4% paraformaldehyde (Sigma, St Louis, MO, USA) for 30 min subsequently washed with PBS. For 5 s, cells were permeabilized using 0.1% Triton-X (Sigma, St Louis, MO, USA) and then washed with PBS. Fixed A549 cells were dyed with DAPI (Sigma, St Louis, MO, USA) and Rhodamine Phalloidin (Molecular Probe, Eugene, OR, USA) for 10 min, washed, and subsequently mounted with glycerol (Sigma, St Louis, MO, USA) for fluorescence imaging. Single-cell fluorescent measurements were made using a microscope (Nikon Eclipse TE 2000-S, Tokyo, Japan).

3. Results

3.1. Characteristics of PP-SCT

As the D is increased, the angle (θ) and the height (h) are also increased. For a standard Petri dish of 90 mm × 15 mm and a 3 × 1-inch microscope glass slide, the scope of operation is shown in Figure 6. When D is increased beyond 50 mm for LD and 60 mm for UD, the θ and h values are increased more significantly. As h is increased, the fluid velocity was increased linearly. The fluid velocity range of around 0.25–4 m/s (a fluid channel of 2 mm width and 100 μm height) were achieved using the PP-SCT technique for the considered design, which can be either increased or decreased depending on the fluid channel design. Figure 7 shows the relationship between height, fluid velocity, and tilt angle. UD can be used for low flow rate applications and LD for high flow rate applications.

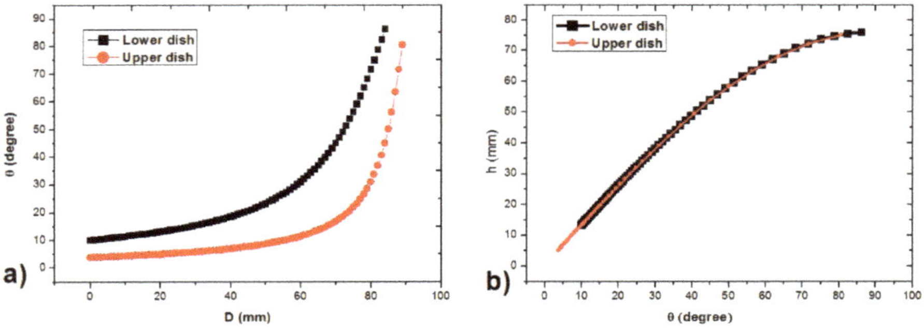

Figure 6. (**a**) Graph plot showing the relationship between D and θ; (**b**) a graph showing the relationship between θ and h. This shows the scope of the operation of the lower and upper dish.

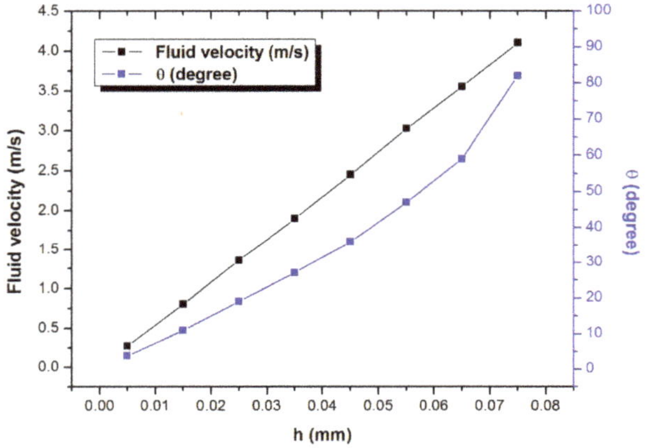

Figure 7. Graph plot showing the relationship between h, θ, and fluid velocity. Fluid velocity was obtained from the simulation study corresponding to the 2 mm fluid channel with a 100 μm height.

3.2. Single-Cell Trapping and Design Analysis

Streamline plots of three microfluidic channel designs are shown in Figure 8. Flow trajectories are varying in the serpent-shaped channel compared to the other two designs. Observations from three biochip designs reveal that the single-cell trapping was greater in the serpent-shaped biochip followed by the branched channel design and the straight channel design. In the serpent design, SCC was greater and MCC and NC (no capture) were lower when compared to the other two designs (refer to Figures 9 and 10).

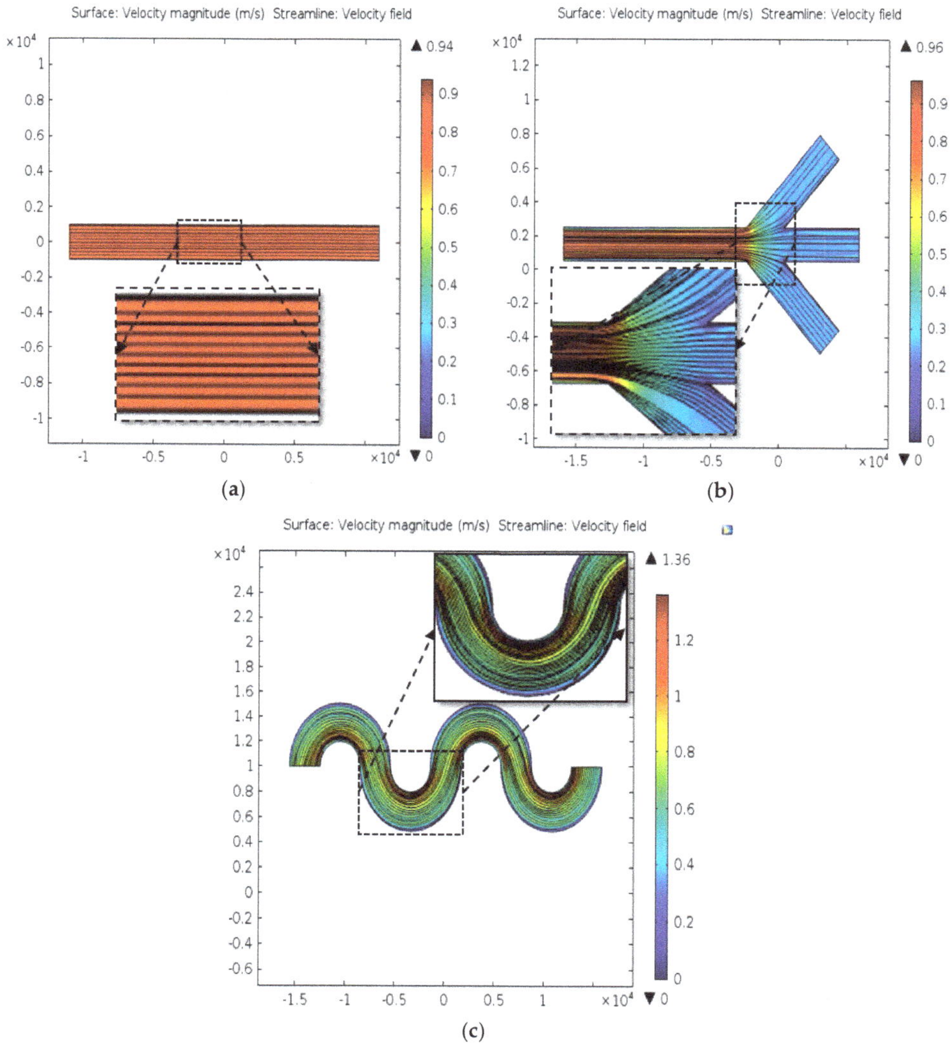

Figure 8. Streamline flow velocity profile in three fluid channels, the inset shows the close-up view of (**a**) the straight channel, (**b**) the branched channel, and (**c**) the serpent-shaped channel.

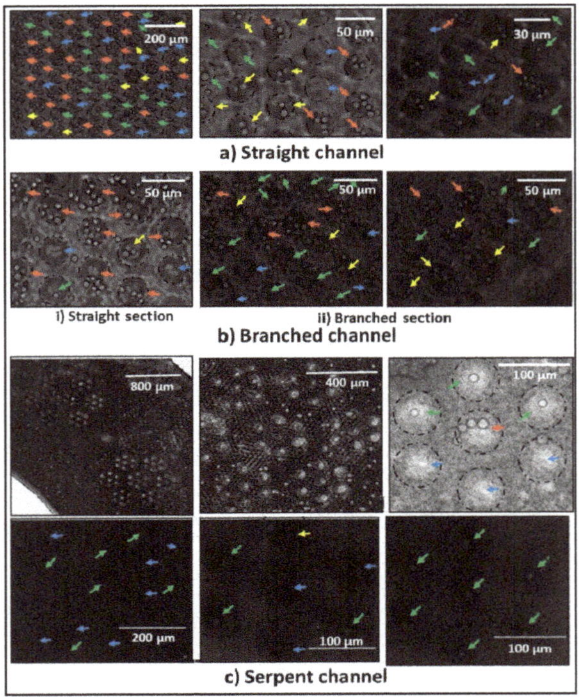

Figure 9. Trapped A549 cells in (**a**) the straight channel design; (**b-i**) the straight section in the branched channel design; (**b-ii**) the side-branched section in the branched channel design, and (**c**) serpent channel design. The blue arrow points to no capture, the green arrow points to SCC, the yellow arrow points to dual-cell capture, and the red arrow points to three or more cell captures. The dotted black circles were just for clarity; it did not resemble the exact scale of the biochip.

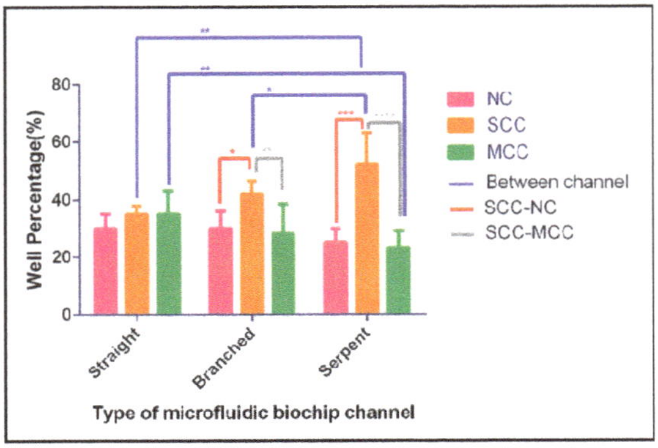

Figure 10. Cell trapping capabilities of three channel designs with the comparison of NC, SCC, and MCC based on the study conducted. Mean and SD were plotted from five independent repeated experiments. ANNOVA analysis was done within each design to find the significant parameters and two-way ANNOVA analysis was done comparing the designs to find the significant parameters. It can be seen that in the serpent design NC and MCC are the least, and SCC is the maximum in comparison to the others.

3.3. Cell Viability and Single-Cell Fluorescent Measurements

It is evidenced by the results that viable A549 cells were trapped in microwells. The cells observed were shining as the cells did not take up trypan blue (refer to Figure 11a–c). If cells were dead, it would take up the trypan blue. Almost no dead cells were observed. Single-cell fluorescence signals were observed in the demonstrated biochip. The results reveal that the A549 cells had intact morphology. The DAPI (blue) signal from the nucleus and the Rhodamine Phalloidin (red) signal from actin filaments, which holds the cytoskeleton of the cells. These results confirm that the cell shows attachment at the bottom of the microwell for short-term study (refer to Figure 11d,e).

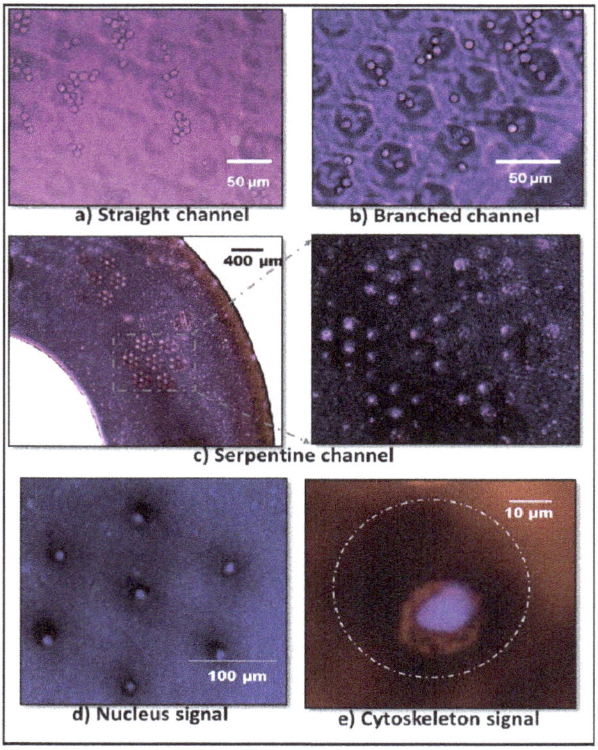

Figure 11. (**a**–**c**) shows the cell viability test (cells did not take up the trypan blue) on the straight channel, branched channel, and serpent channel, respectively; (**d**) the fluorescence signal from the nucleus (DAPI) of single cells being captured in the serpent channel; and (**e**) single-cell cytoskeleton (fluorescence image) after 3 h of incubation from loading to the serpent channel biochip, showing the blue signal from the nucleus (DAPI) and red from the actin (Rhodamine Phalloidin).

4. Discussion

Conventional approaches involve auxiliary systems supporting microfluidic platforms which add to the complexity, lab space, and cost. Mostly the developed approaches involve experienced handling, and some involve application and exposure to electric fields and magnetic fields. The method which does not require experienced handling is in great demand. PP-SCT can be used in the environment where access to any external force or electric is absent. The biochip consists of four parts: an inlet reservoir where the cell suspension is introduced, the main channel, microwells patterned on the bottom surface of the main channel, and the outlet reservoir. Factors that are most likely to affect the cell occupancy in the PP-SCT technique include: (1) the shape of the channel; (2) the size, positioning, and orientation of microwells; (3) the size of the cells; (4) the cell concentration; (5) the biochip tilt angle; and (6) the glass slide length. These factors are to be considered while designing a hydrodynamic-based microfluidic biochip for high SCC. The scope of the PP-SCT is very wide as it is simple, cost effective, and can be readily implemented for any SCA as it does not require any additional equipment. SC-PPT also provides a varying operational range of controlling the fluid flow rate depending on the tilt angle, without involving any auxiliaries. Further, the PP-SCT technique does not harm the cells, which were evidenced by the viability tests. From the results, it is revealed that the geometry of fluid channels affects the trapping rate and efficiency of the SCC rate. There are more possibilities for trapping multiple cells in the straight and branched channel. This is because when cells travel over any regions of the straight or branched channel, they travel nearly the same distance, speed, and trajectory profile to the well, providing high chances of multiple cells being trapped anywhere in the channel. In the branched channel after the branch section, cell trapping was found to be improved. This is due to the varying trajectory profile at the branched section of the branched channel. On the contrary, the serpent channel with a hexagonally-positioned microwell array provides efficient single-cell trapping possibilities. This is due to varying trajectories of the flow in the channel as shown in the simulation results. This enables the cells to travel in the varying trajectory path with less clumping, thereby providing high SCC with the least MCC. Achieving high SCC with no or minimal NC and MCC is great if that demand can be achieved without involving specialized equipment and no auxiliary equipment would definitely add to the upside. Advantages of using vertical cell trapping when compared to lateral trapping are that the cell analytes can be treated or measured individually without clump analyte measurement. In a vertical well, single-cell analytes are preserved separately for analysis as an analogy to the conventional six-well or 96-well plate. Single-cell fluorescent images reveal that cell morphology was intact, which is suitable for short-term investigation of a single cell. This technique enables the real-time observation of single-cell measurements for analysis. Thus, these observations act as design guidelines for the PP-SCT technique for hydrodynamic-based non-auxiliary biochips in achieving high SCC. The PP-SCT technique can be combined with open tools [41] for further exploration of the single cell analysis, making it accessible and cost effective.

5. Conclusions

This article has demonstrated the PP-SCT technique, along with the design, fabrication, and analysis of a micro-fluid dynamic-based biochip for single-cell trapping. Flow rate variations can be precisely achieved using PP-SCT technique with a flow velocity range of 0.25–4 m/s (with a fluid channel of 2 mm width and 100 μm height). Characterization of three fluid channel shapes for improving SCC was performed. Observations reveal that SCC rates can be improved based on channel shape and orientation of the microwell array. SCC was around 52%, 42%, and 35% for the serpent, branched, and straight channel biochips, respectively. In this investigation it is evident that SCC can be improved significantly, based on the microchannel design, for a fixed cell density, with a similar flow rate without involving any auxiliary systems. This article further paves the way for designing nonauxiliary hydrodynamic-based microfluidics for high SCC with no, or minimal, NC and MCC. Single-cell fluorescence measurements were performed with the biochip. The PP-SCT technique investigated has wide applications in the field of single-cell microfluidic-omics technologies, such as genomics, proteomics, secretomics, and metabolomics. This result can help in designing a passive microfluidic biochip with lateral and vertical microwell arrays for efficient single-cell trapping using the PP-SCT technique. Future directions are to characterize the PP-SCT for the different angles with compatible microfluidic designs for high SCC.

Author Contributions: V.N. and F.S. designed the whole project, performed the experiments, analyzed the results, and wrote the paper. T.P.L. and K.M. contributed to the work of biochip fabrication and reviewing the article. A.Y.F.K., H.A.H., and M.B.B. contributed to the work of cell culture and fluorescence imaging.

Funding: The authors would like to thank and acknowledge the Exploratory Research Grant Scheme by the Ministry of Education, Malaysia (ERGS-MOE RDU120606).

Acknowledgments: The PDF scholarship conferred to V.N. by University Malaysia Pahang is gratefully acknowledged.

Conflicts of Interest: The authors declare no conflict of interest.

Abbreviation

PP-SCT	Pipette Petri dish single-cell trapping
SCC	Single-cell capture
MCC	Multiple-cell capture
NC	No capture
UD	Upper dish
LD	Lower dish
SCA	Single cell analysis
PET	Polyethylene terephthalate
IPA	Isopropyl alcohol
CGM	Complete growth media
PBS	Phosphate-buffered solution

References

1. Khoshnoud, F.; de Silva, C.W. Recent advances in mems sensor technology–biomedical applications. *IEEE Instrum. Meas. Mag.* **2012**, *15*, 8–14. [CrossRef]
2. Kim, C.-S.; Ahn, S.-H.; Jang, D.-Y. Review: Developments in micro/nanoscale fabrication by focused ion beams. *Vacuum* **2012**, *86*, 1014–1035. [CrossRef]
3. Vigneswaran, N.; Samsuri, F.; Ranganathan, B. Recent advances in nano patterning and nano imprint lithography for biological applications. *Procedia Eng.* **2014**, *97*, 1387–1398. [CrossRef]
4. Wang, D.; Bodovitz, S. Single cell analysis: The new frontier in 'omics'. *Trends Biotechnol.* **2010**, *28*, 281–290. [CrossRef] [PubMed]
5. Dittrich, P.; Jakubowski, N. Current trends in single cell analysis. *Anal. Bioanal. Chem.* **2014**, *406*, 6957–6961. [CrossRef] [PubMed]
6. Navin, N.; Hicks, J. Future medical applications of single-cell sequencing in cancer. *Genome Med.* **2011**, *3*, 31. [CrossRef] [PubMed]
7. Wang, Y.; Navin, N.E. Advances and applications of single-cell sequencing technologies. *Mol. Cell* **2015**, *58*, 598–609. [CrossRef] [PubMed]
8. Occhetta, P.; Malloggi, C.; Gazaneo, A.; Redaelli, A.; Candiani, G.; Rasponi, M. High-throughput microfluidic platform for adherent single cells non-viral gene delivery. *RSC Adv.* **2015**, *5*, 5087–5095. [CrossRef]
9. Fujita, H.; Esaki, T.; Masujima, T.; Hotta, A.; Kim, S.H.; Noji, H.; Watanabe, T.M. Comprehensive chemical secretory measurement of single cells trapped in a micro-droplet array with mass spectrometry. *RSC Adv.* **2015**, *5*, 16968–16971. [CrossRef]
10. Sun, H.; Olsen, T.; Zhu, J.; Tao, J.; Ponnaiya, B.; Amundson, S.A.; Brenner, D.J.; Lin, Q. A bead-based microfluidic approach to integrated single-cell gene expression analysis by quantitative rt-pcr. *RSC Adv.* **2015**, *5*, 4886–4893. [CrossRef] [PubMed]
11. Whitesides, G.M. The origins and the future of microfluidics. *Nature* **2006**, *442*, 368–373. [CrossRef] [PubMed]
12. Rettig, J.R.; Folch, A. Large-scale single-cell trapping and imaging using microwell arrays. *Anal. Chem.* **2005**, *77*, 5628–5634. [CrossRef] [PubMed]
13. Di Carlo, D.; Wu, L.Y.; Lee, L.P. Dynamic single cell culture array. *Lab Chip* **2006**, *6*, 1445–1449. [CrossRef] [PubMed]
14. Park, J.Y.; Morgan, M.; Sachs, A.N.; Samorezov, J.; Teller, R.; Shen, Y.; Pienta, K.J.; Takayama, S. Single cell trapping in larger microwells capable of supporting cell spreading and proliferation. *Microfluid. Nanofluid.* **2010**, *8*, 263–268. [CrossRef] [PubMed]
15. Kobel, S.; Valero, A.; Latt, J.; Renaud, P.; Lutolf, M. Optimization of microfluidic single cell trapping for long-term on-chip culture. *Lab Chip* **2010**, *10*, 857–863. [CrossRef] [PubMed]

16. Wang, Y.; Shah, P.; Phillips, C.; Sims, C.E.; Allbritton, N.L. Trapping cells on a stretchable microwell array for single-cell analysis. *Anal. Bioanal. Chem.* **2012**, *402*, 1065–1072. [CrossRef] [PubMed]
17. Banaeiyan, A.A.; Ahmadpour, D.; Adiels, C.B.; Goksör, M. Hydrodynamic cell trapping for high throughput single-cell applications. *Micromachines* **2013**, *4*, 414–430. [CrossRef]
18. Huang, L.; Chen, Y.; Chen, Y.; Wu, H. Centrifugation-assisted single-cell trapping in a truncated cone-shaped microwell array chip for the real-time observation of cellular apoptosis. *Anal. Chem.* **2015**, *87*, 12169–12176. [CrossRef] [PubMed]
19. Ashok, P.C.; Dholakia, K. Optical trapping for analytical biotechnology. *Curr. Opin. Biotechnol.* **2012**, *23*, 16–21. [CrossRef] [PubMed]
20. Wu, M.; Singh, A.K. Single-cell protein analysis. *Curr. Opin. Biotechnol.* **2012**, *23*, 83–88. [CrossRef] [PubMed]
21. Li, M.; Xu, J.; Romero-Gonzalez, M.; Banwart, S.A.; Huang, W.E. Single cell raman spectroscopy for cell sorting and imaging. *Curr. Opin. Biotechnol.* **2012**, *23*, 56–63. [CrossRef] [PubMed]
22. Lee, G.-H.; Kim, S.-H.; Ahn, K.; Lee, S.-H.; Park, J.Y. Separation and sorting of cells in microsystems using physical principles. *J. Micromech. Microeng.* **2015**, *26*, 013003. [CrossRef]
23. Johann, R.M. Cell trapping in microfluidic chips. *Anal. Bioanal. Chem.* **2006**, *385*, 408–412. [CrossRef] [PubMed]
24. Nilsson, J.; Evander, M.; Hammarström, B.; Laurell, T. Review of cell and particle trapping in microfluidic systems. *Anal. Chim. Acta* **2009**, *649*, 141–157. [CrossRef] [PubMed]
25. Lo, S.-J.; Yao, D.-J. Get to understand more from single-cells: Current studies of microfluidic-based techniques for single-cell analysis. *Int. J. Mol. Sci.* **2015**, *16*, 16763–16777. [CrossRef] [PubMed]
26. Valizadeh, A.; Khosroushahi, A.Y. Single-cell analysis based on lab on a chip fluidic system. *Anal. Methods* **2015**, *7*, 8524–8533. [CrossRef]
27. Shi, Q.; Qin, L.; Wei, W.; Geng, F.; Fan, R.; Shin, Y.S.; Guo, D.; Hood, L.; Mischel, P.S.; Heath, J.R. Single-cell proteomic chip for profiling intracellular signaling pathways in single tumor cells. *Proc. Natl. Acad. Sci. USA* **2012**, *109*, 419–424. [CrossRef] [PubMed]
28. Kim, H.S.; Devarenne, T.P.; Han, A. A high-throughput microfluidic single-cell screening platform capable of selective cell extraction. *Lab Chip* **2015**, *15*, 2467–2475. [CrossRef] [PubMed]
29. Eyer, K.; Kuhn, P.; Hanke, C.; Dittrich, P.S. A microchamber array for single cell isolation and analysis of intracellular biomolecules. *Lab Chip* **2012**, *12*, 765–772. [CrossRef] [PubMed]
30. Shirai, M.; Taniguchi, T.; Kambara, H. Emerging applications of single-cell diagnostics. *Top. Curr. Chem.* **2012**, *336*, 99–116.
31. Yi, C.; Li, C.-W.; Ji, S.; Yang, M. Microfluidics technology for manipulation and analysis of biological cells. *Anal. Chim. Acta* **2006**, *560*, 1–23. [CrossRef]
32. Kim, S.M.; Lee, S.H.; Suh, K.Y. Cell research with physically modified microfluidic channels: A review. *Lab Chip* **2008**, *8*, 1015–1023. [CrossRef] [PubMed]
33. Bhagat, A.A.S.; Bow, H.; Hou, H.W.; Tan, S.J.; Han, J.; Lim, C.T. Microfluidics for cell separation. *Med. Biol. Eng. Comput.* **2010**, *48*, 999–1014. [CrossRef] [PubMed]
34. Burger, R.; Ducrée, J. Handling and analysis of cells and bioparticles on centrifugal microfluidic platforms. *Expert Rev. Mol. Diagn.* **2012**, *12*, 407–421. [CrossRef] [PubMed]
35. Cetin, B.; Özer, M.B.; Solmaz, M.E. Microfluidic bio-particle manipulation for biotechnology. *Biochem. Eng. J.* **2014**, *92*, 63–82. [CrossRef]
36. Karimi, A.; Yazdi, S.; Ardekani, A. Hydrodynamic mechanisms of cell and particle trapping in microfluidics. *Biomicrofluidics* **2013**, *7*, 021501. [CrossRef] [PubMed]
37. Batchelor, G.K. *An Introduction to Fluid Dynamics*; Cambridge University Press: Cambridge, UK, 1967; Volume 515, p. 13.
38. Stokes, G.G. *On the Effect of the Internal Friction of Fluids on the Motion of Pendulums*; Pitt Press: Pittsburgh, PA, USA, 1851; Volume 9.
39. Reynolds, O. An experimental investigation of the circumstances which determine whether the motion of water shall be direct or sinuous, and of the law of resistance in parallel channels. *Proc. R. Soc. Lond.* **1883**, *35*, 84–99. [CrossRef]

40. Lee, T.P.; Mohamed, K. *3D Microfabrication Using Emulsion Mask Grayscale Photolithography Technique*; IOP Conference Series: Materials Science and Engineering; IOP Publishing: Bristol, UK, 2016; p. 012032.
41. Kuznetsov, S.; Doonan, C.; Wilson, N.; Mohan, S.; Hudson, S.E.; Paulos, E. Diybio things: Open source biology tools as platforms for hybrid knowledge production and scientific participation. In Proceedings of the 33rd Annual ACM Conference on Human Factors in Computing Systems, Seoul, Korea, 18–23 April 2015; ACM: New York, NY, USA, 2015; pp. 4065–4068.

fluids

MDPI

Article

Advancing Chemical Risk Assessment through Human Physiology-Based Biochemical Process Modeling

Dimosthenis Sarigiannis [1,2,3,*] and Spyros Karakitsios [1,2,3]

[1] HERACLES Research Center on the Exposome and Health, Center for Interdisciplinary Research and Innovation, Balkan Center, Bldg. B, 10th km Thessaloniki-Thermi Road, 57001 Thermi, Greece; spyros.karakitsios@gmail.com

[2] Environmental Engineering Laboratory, Department of Chemical Engineering, Aristotle University of Thessaloniki, University Campus, 54124 Thessaloniki, Greece

[3] Environmental Health Engineering, Department of Science, Technology and Society, University School for Advanced Study (IUSS), Piazza della Vittoria 15, 27100 Pavia, Italy

* Correspondence: denis@eng.auth.gr; Tel.: +30-2310-99-4562

Received: 30 September 2018; Accepted: 27 December 2018; Published: 4 January 2019

Abstract: Physiology-Based BioKinetic (PBBK) models are of increasing interest in modern risk assessment, providing quantitative information regarding the absorption, metabolism, distribution, and excretion (ADME). They focus on the estimation of the effective dose at target sites, aiming at the identification of xenobiotic levels that are able to result in perturbations to the biological pathway that are potentially associated with adverse outcomes. The current study aims at the development of a lifetime PBBK model that covers a large chemical space, coupled with a framework for human biomonitoring (HBM) data assimilation. The methodology developed herein was demonstrated in the case of bisphenol A (BPA), where exposure analysis was based on European HBM data. Based on our calculations, it was found that current exposure levels in Europe are below the temporary Tolerable Daily Intake (t-TDI) of 4 μg/kg_bw/day proposed by the European Food Safety Authority (EFSA). Taking into account age-dependent bioavailability differences, internal exposure was estimated and compared with the biologically effective dose (BED) resulting from translating the EFSA temporary total daily intake (t-TDI) into equivalent internal dose and an alternative internal exposure reference value, namely biological pathway altering dose (BPAD); the use of such a refined exposure metric, showed that environmentally relevant exposure levels are below the concentrations associated with the activation of biological pathways relevant to toxicity based on High Throughput Screening (HTS) in vitro studies.

Keywords: biochemical processes; biokinetics; human biomonitoring; bisphenol A; exposure reconstruction; risk assessment

1. Introduction

One of the main applications of internal dosimetry is the integration of exposure and human biomonitoring (HBM) data. More in detail, internal dosimetry aims at (i) deriving the time course of the toxicants in human tissues, with a particular focus on susceptible developmental stages; (ii) providing a comprehensive interpretation of the HBM data related to the cohorts, for quantifying individual exposome; and (iii) deriving Biologically Effective Dose (BED) values for associating them with adverse outcomes. Towards these aims, a lifetime generic Physiology-Based BioKinetic (PBBK) model [1] has been developed, that includes the interaction of chemical mixtures [2], and a set of exposure reconstruction algorithms, starting from HBM data [3]. To expand the chemical space that is covered by the generic PBBK model, advanced quantitative structure–activity relationship (QSARs) models have been introduced for its parameterization regarding data-poor chemicals. For reconstructing exposure based on HBM data [4], a multiple tier methodology has been developed, that accounts for the availability of data (meaning the number of samples collected for capturing intra-individual variability), as well as the needs of the exposure assessment (e.g., the daily intake estimation of exposure peaks during the day). Thus, several methods are available, including the Exposure Conversion Factors (ECFs) [5] and the Markov Chain Monte Carlo analysis. PBBK models are also applicable in the estimation of xenobiotics levels in target tissues, relevant to the activation of toxicity pathways [6], possibly associated with adverse outcomes. The advent of -omics technologies and their link to BED could highlight the early identification of disease onset. Moreover, BED could be used for the quantification of the effect of extracellular perturbations on metabolic states, coupling the PBBK model with metabolic regulatory networks and defining the feedback loop that connects clearance and metabolite production rates to metabolism regulation [7] via dynamic flux balance analysis (FBA) [8].

The need for developing generic PBBK models is of great importance in modern risk assessment, especially for compounds where toxicokinetics play a particular role in their overall adverse effects in humans. Among this compounds, Bisphenol A (BPA), is a commonly used plasticizer of increased scientific and regulatory interest. BPA is used in the manufacture of polycarbonate plastics and epoxy resins, which were extensively used in baby bottles (its use is now banned) as protective coatings on food containers, as well as in dentistry; other applications of BPA include thermal paper and polyvinyl chloride industries [9]. BPA is has been characterized as an endocrine-disrupting chemical (EDC). It has been suggested that in utero exposure to BPA may disrupts early life development. Many in vitro and in vivo (animal) studies demonstrated that adverse health effects due to BPA exposure can occur also at environmentally relevant doses [10]. However, it has to be noted that these results are highly controversial, since numerous high-quality attempts to replicate these in vivo at "environmentally relevant doses" have not succeeded [11–14]. BPA is one of the compounds with the highest controversy regarding its toxicokinetics; although BPA glucuronidation (which is the major elimination pathway) is very rapid at early developmental stages, internal exposure is higher as a result of the immature detoxification process [15,16].

Given all the above, this study aims at providing some answers on the risks associated with BPA exposure of the European population, starting from HBM data. Towards this aim, (a) a lifetime PBBK model has been developed, that accounts for the early developmental stages (including in utero exposure), and at the same time covers a large chemical space and (b) HBM data are assimilated with an exposure reconstruction mathematical framework, coupled with the PBBK model. It has also to be noted, that the PBBK model incorporates the key routes of exposure namely inhalation, oral and dermal), and includes multiple compartments, as well as binding to plasma proteins and red blood cell binding. At the same time, the model is able to simulate the levels of the parent compounds and their metabolites in the various tissues and biological fluids (including maternal milk). It also has to be noted that the assessment of internal dosimetry using PBBK models allows for the calculation of the xenobiotics internal doses, above the thresholds associated with biological pathway alterations that might be relevant to adverse outcomes. The so-called biological pathway altering dose (BPAD) provides additional advantages compared to existing risk metrics, since it combines dose—response data with the analysis of uncertainty and population variability, so as to derive exposure limits and can be derived by high-throughput screening (HTS) in vitro data. The overall applicability of the developed PBBK model in advancing the risk assessment concepts, is demonstrated in the case of BPA.

2. Materials and Methods

2.1. Development of the Generic Lifetime PBPK Model

The physiology-based biokinetic model developed and outlined in this work is designed to take into account the evolution of enzymatic activity and availability, as well as a change in organ volumes and blood flow characteristics during human life. These attributes determine the overall ADME properties of the human body, allowing the model to cover a large chemical space, if appropriately parameterised. The mathematical formulation adopted describes ADME processes for a wide spectrum of xenobiotics and for their metabolites, reaching all the way down to three metabolic stages [17]. The model concept is shown in Figure 1.

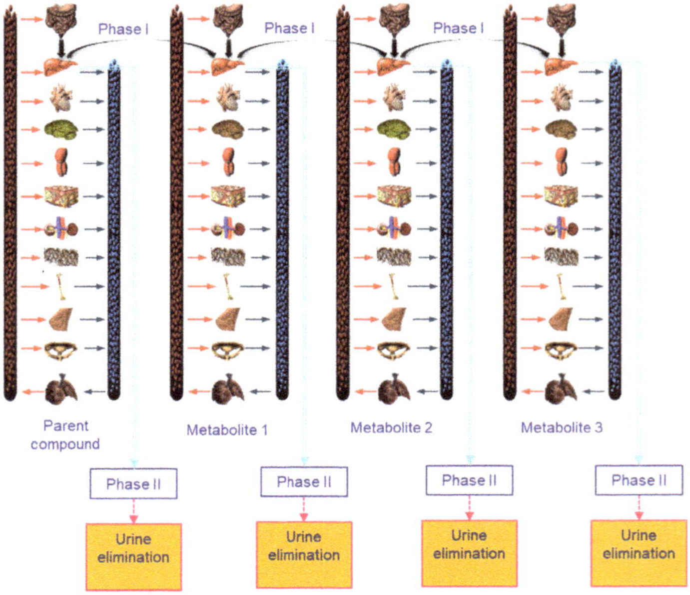

Figure 1. Conceptual representation of the generic Physiology-Based BioKinetic (PBBK) model.

A key problem with generic PBBK models is the need to quantitatively estimate the values of a large number of physiochemical (e.g., the octanol–water partition coefficient, kow), and biochemical (e.g., V_{max} and K_m) parameters corresponding to a large number of physiological and anatomic compartments and biological fluid fluxes. Conventionally, a prohibitively large amount of experiments in vivo and, possibly, in vitro would be required to perform the proper model parameterization, while inter-species extrapolation in the values of these parameters introduces a significant source of uncertainty in the model set-up and, ultimately, its results. Our work has shown that it is feasible to use quantitative structure–activity relationships (QSARs) to estimate model parameters to ensure that the model performs well for a wide array of chemicals. Abraham's solvation equation was used to estimate such biological/biochemical properties. This QSAR formulation takes into account molecular attributes such as:

- excess molar refraction; a property that can be determined if the compound refractive index is known,
- compound dipolarity/polarizability,
- solute effective or summation hydrogen-bond acidity,
- solute effective or summation hydrogen-bond basicity, and
- McGowan characteristic volume that can be calculated based on the molecular structure of the solute.

Coupling Abraham's solvation equation with machine learning algorithms such as feed backward artificial neural networks gives excellent results in terms of predicting chemical-specific biological/biochemical properties such as blood-tissue partition coefficients, maximal velocity (V_{max})

and the Michaelis—Menten constant. In fact the predictive capacity of the QSAR regarding the Michaelis—Menten constant moved from rather poor (R^2 up to 0.35) to an R^2 of 0.88 for 85 chemical substances [18], while the predictive capacity of the model partition coefficients and vmax was very high (R^2 above 0.9) [18,19]. QSAR predictions of these fundamental parameters are given for all major organs, as well as for arterial, venous, and portal blood. Metabolically active tissues link up xenobiotic compounds and their metabolites. Such tissues include mainly the liver, but also other sites of metabolism such as the intestine, the brain, the skin, or the placenta based on the presence or not of the respective enzymes that catalyze the metabolism of the specific chemical substance. Mass balance in each body compartment describes all relevant processes to absorption, metabolism, elimination, and protein binding. In each tissue, three mass balance equations are written, for (a) red blood cells, (b) plasma and interstitial tissue, and (c) cells. Thus, the model is readily applicable to both flow- and membrane-limited chemicals. Specific organs are further divided in sub-compartments to account for inhomogeneity in enzymatic activity (e.g., the liver is divided in five zones to describe the spatial distribution of enzymes), and permeability differences (e.g., the brain is divided in four sub-compartments, i.e., main brain, globus palidus, cerebellum and pituitary gland).

Our model captures in utero exposure by describing the interactions between the pregnant mother and her fetus by way of modelling the intra-placental properties that determine chemical transfer from the mother to the fetus during the different phases of growth of the latter in utero. Anthropometric parameters used in the model depend on age, so as to support the estimation of internal dose over the lifecourse, and to capture the continuously changing physiology of both the mother and the developing embryo. Diffusion is the main process governing flow from uterus to the placenta and vice-versa during pregnancy [20]. Excretion via lactation is described as an output from mammary tissue through a partitioning process between mammary tissue and milk, and milk withdrawal by suckling. The mathematical formulation describing lactation is based on the one described for PCBs in rats [21], which was adopted for humans [22]. The three main exposure routes are explicitly distinguished in the model set up. Inhalation takes into account absorption of gases and deposition of particles size fractions across the human respiratory tract. Absorption through the oral route is governed by the absorption rates of the stomach and the gut. To better describe dermal absorption, the skin was modelled as a multi-layer structure, including a brick-and-mortar model of the stratum corneum [23] and viable epidermis (also accounting for metabolism), where the geometry of all layer microstructures was explicitly described [24]. Details on the model mathematical framework are given in the Supplementary Materials.

2.2. BPA Toxicokinetic Considerations

In this study, the comprehensive PBBK model developed by Sarigiannis et al. [1,25] for BPA was used. The model describes in detail the ADME processes related to BPA and its conjugated metabolite, BPA-Glu. The latter was considered in particular, because the main detoxification pathway of BPA is phase II glucuronidation. The original model was scaled for children using the physiologically-based approach proposed by Edginton et al. [26], and data about ontogeny of enzymes (glycotransferases) involved in BPA detoxification [27,28]. Following a method similar to Edginton and Ritter [29], plasma clearance (CL_H) from adults was converted to intrinsic clearance (Clint) using the well-stirred reactor model, physiological data regarding liver blood flow (Q_H; mL/min) and the unbound fraction of BPA in the plasma (f_u). Elimination through urine (in the form of BPA-Glu) is the only mechanism of elimination. We optimized plasma $CL_{BPA\text{-}Glu_}$kidney during the development of the adult coupled model and compared this with values derived from Volkel et al. [29]. It has to be noted, that this parameterization of our model, resulted in similar levels of plasma BPA and urinary BPA-Glu to the ones derived by Thayer et al. [30] and Teeguarden et al. [31]; to our knowledge, these are the only studies where BPA has been administered to volunteers and data useful for PBPK modeling have been derived; however, it has to be noticed that in order to fit in the optimal way the measured serum concentrations, an adjustment of the oral uptake constant is required, as explicitly described by Yang et al. [32]. Tissue:blood partition coefficients were obtained from Edginton and Ritter [15]. Other age- and gender-specific data on parameters, such as urinary volumes and creatinine excretion rate were obtained from the International Commission on Radiological Protection [33]. Global sensitivity analysis was performed by changing the input parameters by 1%, while observing the relative change in the outcome. Results showed that the rapid metabolism and elimination of BPA, as well as its strong binding to plasma proteins, were the parameters affecting the most model performance.

2.3. Exposure Reconstruction Starting from Human Biomonitoring (HBM) Data

Reverse modeling algorithms applied on the PBBK model allowed us to reconstruct exposure from human biomonitoring (HBM) data. The comprehensive interpretation of human biomonitoring data and their back-calculation into exposure distributions comprises a typical computational inversion problem. The aim is to estimate input distributions that explain the measured biomonitored data, and at the same time to minimize the residual error. Towards this aim, ancillary exposure information is needed, related to environmental contamination, and human activities (including diet and consumer products use) [34]. An iterative computational methodology was developed based on Bayesian Markov Chain Monte Carlo (MCMC), combined with the generic PBBK model aiming at performing accurate exposure reconstruction. Differential Evolution (DE) and MCMC algorithms have been combined to solve this problem for the first time. The PBBK model has been combined with the Bayesian MCMC [35,36] and differential evolution Monte Carlo (DEMC) [37] techniques in order to simulate and calculate the exposure value that fits best the observed HBM data.

2.4. Exposure Assessment

The most effective way to estimate exposure and intake is the use of human biomonitoring data. For the assessment of BPA exposure, HBM data were collected from the available literature. Urinary BPA concentrations reflect differences in consumer exposure that are related to food packaging material (canned food, milk formula, use of plastic baby bottles). The restriction of BPA use in baby bottles in the European Union Member States in 2011, and increasing public awareness regarding the potential adverse health outcomes related to BPA exposure resulted in a decline of measured BPA levels in the European population over the last eight years. In the most recent studies, urinary BPA (in the form of the glucuronidated metabolite) measured levels are about 2 μg/L. BPA urinary concentrations measured in Germany (German Environmental Survey—GerES) depend on child age: they were 3.5, 2.8, 2.1 and 2.6 μg/L for children aged 3–5, 6–8, 9–11 and 12–14 years old respectively. Similar levels (2.5 μg/L) were recorded in France [38] for pregnant women. According to the German ESB [39] urinary concentrations of BPA declined from 2 μg/L in 1995 to 1.3 μg/L in 2009. The multi-center study DEMOCOPHES [40] provided results for urinary BPA levels in Belgium (2.6 μg/L), Denmark (2.2 μg/L), Luxembourg (1.9 μg/L), Slovenia (2.1 μg/L), Spain (2.1 μg/L) and Sweden (1.4 μg/L) [41]. The results of the urinary BPA levels in various countries are summarized in Table 1.

Table 1. Urinary bisphenol A (BPA)-Glu levels from European human biomonitoring studies.

Country—Study Name	Population Group	Mean	Median	Reference
Belgium—Democophes	Mothers (≤45 years)		2.6	[41]
Denmark—Democophes	Mothers (≤45 years)		2.2	
Denmark—Copenhagen Puberty Study	Children and adolescents (5–9 years)	2.3		[42]
	Children and adolescents (10–13 years)	1.5		
	Children and adolescents (14–20 years)	0.7		
Denmark—Copenhagen Study on Male Reproductive Health	Young men	3.2		
Denmark—Odense Child Cohort	Pregnant women	1.5		

Table 1. *Cont.*

Country—Study Name	Population Group	Mean	Median	Reference
France—ELFE	Pregnant women	2.5	2	[38]
Germany—ESB	Students (<2000)—Münster Students (≥2000)—Münster		2.0 1.4	[39]
Germany—GerES	3–14 years 3–5 years 6–8 years 9–11 years 12–14 years	2.7 3.5 2.8 2.1 2.6	2.7 3.6 2.7 2.2 2.4	[43]
Italy—InCHIANTI	20–40 years 41–65 years 66–74 years	4.4 3.9 3.3	4.3 3.7 3.2	[44]
Luxembourg—Democophes	Mothers (≤45 years)		1.9	[41]
Netherlands—Generation R	Pregnant women (18–41 years)	1.2	1.1	[45]
Slovenia—Democophes	Mothers (≤45 years)		1.2	[41]
Spain—INMA Spain—INMA	Pregnant women Children (4 years)	2.2 4.2		[46]
Spain—Democophes Sweden—Democophes	Mothers (≤45 years) Mothers (≤45 years)		2.1 1.4	[41]
France—ELFE	Pregnant women (18–40 years)	0.7		[47]
Greece—Rhea	Pregnant women Children (2 years)	1.2 2.0	1.2 2.1	[48]

2.5. *Risk Assessment*

A risk characterization of BPA was carried out, employing several tools related to external and internal exposure assessment. To start with the value set as tolerable daily intake (t-TDI) by European Food Safety Authority (EFSA) temporarily 4 μg/kg_bw/day was used [49]. The options for evaluating exposure levels included:

a. Direct comparison of exposure reconstruction intake estimates to EFSA t-TDI of 4 μg/kg_bw/day.
b. Use of a biomonitoring equivalent (BE) value for urinary data. An original BE for BPA has been derived by Krishnan et al. [50] equal to 2000 μg/L, on the basis of the old EFSA TDI (equal to 50 μg/kg_bw/day), following the original BE concept initially proposed by Hays et al. [51] and further expanded by Aylward [52]. The reference dose for deriving the BE value was the EFSA t-TDI of 4 μg/kg_bw/day. It was assumed that this dose is given orally to an adult of 70 kg body weight at a constant rate during the day. After that, this intake was fed to the PBBK model resulting to urinary BPA-Glu concentration of 280 μg/L.
c. Given the limitations of exposure back-calculation based on urinary BPA-Glu levels, the use of another exposure metric more relevant to where the xenobiotics exert their toxicity has to be considered. Towards this aim, free plasma BPA was selected as a descriptive metric linked to the biologically effective dose (BED). The use of this internal exposure metric, allows us to further differentiate internal and external exposure as a result of bioavailability differences related to developmental stage, point of entrance and eventually genetics. As a result, the calculated area under the curve (AUC) for 24 h, equals 0.312 μg 24 h/L (for one hour time interval) [25].

3. Results

3.1. *Exposure Reconstruction based on HBM Data*

The use of the PBBK model coupled to the algorithm for exposure reconstruction resulted in the estimation of daily intake starting from HBM data. In turn, the intake estimates were used as an input in the PBBK model for calculating the internal dose. To estimate the daily intake, some assumptions were done relevant to the exposure scenario, including (a) the average urinary BPA-Glu concentration for each age group of the various countries, and (b) the development of a basic daily schedule of diet that includes three different meals: (a) breakfast at 7:00 a.m. (dose 1), (b) lunch at 2:00 p.m. (dose 2) and (c) dinner at 7:00 p.m. (dose 3). Overall daily intake was considered to be equally distributed among these meals.

The exposure reconstruction algorithm converged to the available biomonitoring data after 1000 iterations and the average intake estimates were 0.05 μg/kg_bw/day, which is far below the EFSA t-TDI (Figure 2). The same occurs for the population exposed to the higher levels of BPA (4.4 μg/L for the 20–40 years old Italian adults, as indicated in Table 1), were the maximum intake levels were found equal to 0.77 Italy μg/kg_bw/day.

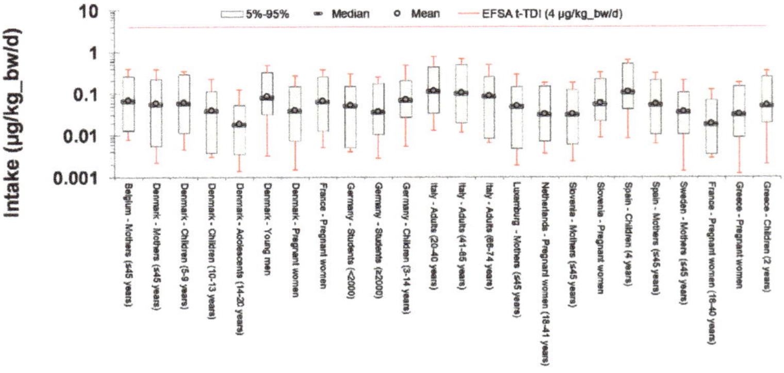

Figure 2. Exposure reconstruction levels.

It has to be noted, that the methodology applied herein, accounts for the urinary metabolites time course. This is a major advantage for reconstructing exposure, because the actual kinetics and the time course of both parent compounds in blood, and of the urinary biomarker measured in urine are captured and described. Thus, using the calculated uptake levels, internal dose is estimated by running forward the model. This is typically represented by the free plasma BPA. The time course of internal exposure, optimally described by area under the curve (AUC) and urinary BPA levels under a typical daily exposure scenario for an average child (a daily uptake of 0.035 μg/kg_bw/day) are graphically illustrated in Figure 3. Internal dose remains low, in the range of ng/L, while the respective AUC (indicated as the shades area under the red line) is in the range of 1.5 ng 24 h/L, which is significantly lower than the one of 0.312 μg 24 h/L, corresponding to a daily intake that is equal to the t-TDI proposed by EFSA.

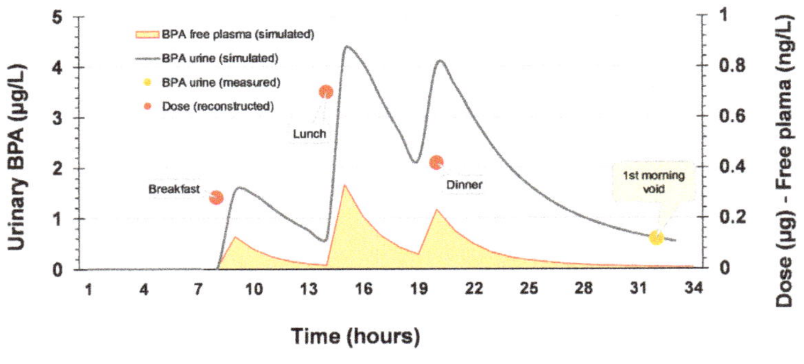

Figure 3. External exposure, internal exposure (area under the curve, AUC), and urinary BPA levels per a typical 24 h exposure scenario. AUC is denoted as the shaded area included between the red line and the x-axis.

3.2. Risk Characterization

Based on the external and internal exposure estimates described above, as well as on the biomonitoring data itself, risk characterization ratios (RCR) were derived for the different age groups. Independently of the RCR method, the median risk for the European population is low (RCR of about 0.01), as illustrated in Figure 4.

For all individuals (including the ones in the upper part of the distribution), RCR < 1, independently of the employed method. It has to be noted that had RCR been estimated solely based on the urinary concentrations, the estimated risks would be lower. In practice, this is an artifact of the comparison between (a) a steady-state calculated urinary concentration, with (b) a urinary concentration that is estimated dynamically in time, accounting for the continuously declining levels of urine eliminated during the night, and collected in the first morning void (Figure 2). On the other hand, the use of internal dose metrics, did not differentiate the RCR results significantly compared to the external intake results. This is somehow expected, because the former, has been calculated based on intake estimates from individual biomonitoring data, which have already accounted for the individual physiology parameters that are needed for the PBBK model. By using the BPAD as a reference value for the internal dose, it was found that even in the worst case scenario, current exposure levels are more than 10 times lower than the $BPADL_{99}$, indicating that that there is no reason for concern regarding BPA exposure.

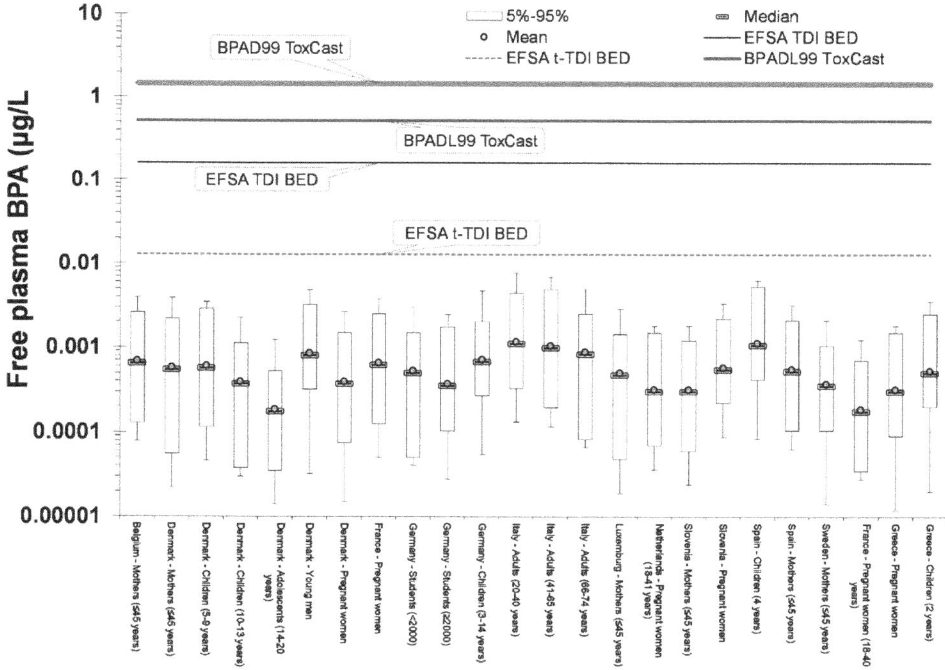

Figure 4. Free plasma BPA for the population included in the included HBM studies (refined internal exposure analysis).

4. Discussion

The study herein deals with the assessment of risks associated with European population with exposure to BPA. In order to estimate the daily intake, an exposure reconstruction scheme has been followed, starting from HBM data. Towards this aim, BPA HBM data has been collated, to fed into a validated PBBK model. After that, the intake estimates were used as an input for the PBBK model for estimating the internal dose (free plasma concentration). In addition, in order to further exploit the opportunities for of internal dosimetry for deriving refined exposure estimates, as well as to better associate urinary data with exposure. Towards this aim, the EFSA t-TDI was translated into a) an equivalent urinary concentration, as well as b) an equivalent internal dose (free plasma concentration).

Based on the urinary levels of morning void and using the exposure reconstruction framework described above, daily intake was estimated. Median intake levels are in the range of 0.05 μg/kg_bw/day, in accordance with recent studies for BPA daily intake that used bottom-up approaches. Daily intake from main dietary sources such as canned food have been estimated at levels of 0.015 μg/kg_bw/day in Belgium [53], while intake from commonly food items and beverages for Norway was 0.005 μg/kg_bw/day [54]. Intake estimates for adults by aggregating the contribution from multiple food items was estimated equal to 0.030 μg/kg_bw/day for Europe [55], while similar results have also been reported for Canada [56] and USA [57].

Beyond these bottom-up approaches, several studies have approximated BPA daily intake, using urinary BPA levels. Considering the very rapid metabolism of BPA, it is assumed that a 24 h urine sample approximates the BPA intake from the previous 24 h [58]. Based on the above assumption, the median daily BPA intake for the US population was estimated at 0.025 μg/kg_bw/day [59]. However, intake estimates based on BPA daily mass balance are representative, only when the whole-day urine is collected. When this is applied in spot samples, the overall intake might be seriously misinterpreted. On the contrary, the use of an exposure reconstruction algorithm, allows the correct interpretation of a spot sample, since exposure and elimination time dynamics are properly described (as already shown in Figure 3).

Another major advantage of exposure reconstruction, is the capability to re-run forward the PBBK model, accounting for all potential parameters that induce bioavailability differences. In a nutshell, these include age, the route of exposure, and genetic variability. Age is affecting the ADME process by differentiating (a) parameters related to human physiology, such as the composition of the various tissues, the blood flows, as well as the inhalation rate, and (b) the maturity of the detoxification pathways. Scaling clearance for children [26] and accounting for the enzymes ontogeny relevant to BPA glucuronidation [27,28,60], an age-dependent bioavailability difference factor of 2 to 3.5 is calculated between neonates and adults. Moreover, significant route-dependent bioavailability differences have been calculated; in the case of inhalation, BPA enters directly in the systemic circulation from the alveoli and the lack of first pass metabolism results in bioavailability differences up to six times, while in the case of dermal exposure, the lack of first-pass metabolism is somehow decompensated by slower absorption. Finally, one of the key factors affecting bioavailability and that is related to inter-individual variability is the existence of genetic variants of key enzymes that are involved in BPA glucuronidation; in vitro kinetic studies have identified that D85Y substitution in UGT2B15 decreases enzymatic function [61], and that the polymorphic alleles of UGT2B15 are translated in variations in the metabolism of BPA [62]. To make these more explicit, the bioavailability differences in free plasma BPA related to developmental stage, point of entrance in human body and genetic polymorphisms of enzymes relevant to metabolism under a daily intake of 4 μg/kg_bw/day are presented in Figure 5. From the above, it is easily concluded that individual variability could be greatly enhanced by the co-existence of factors that predispose for larger bioavailability differences such as early infancy and the presence of the polymorphic alleles of UGT2B15 related to slower metabolism. Should this additional information had been detailed in all the cohorts, larger differences in the internal dose, and also among the RCR assessment would be reported. Thus, it is expected that future cohort studies that properly account for differences in BED (and not only for differences in the biomonitored data), would provide more robust associations between BPA exposure levels and potential health effects.

It has also to be noted, that at the moment, due to the serious analytical limitations, free plasma BPA has been measured only in a very limited number of studies [63–65]. However, it is very encouraging that the estimates of the free plasma concentration of our model (in the range of 0.18 to

1 ng/L), are very much in accordance to the biomonitored doses (in the range of 0.23 to 1.3 ng/L) by Teeguarden et al. [63] for similar urinary BPA-Glu levels (0.9 μg/L) to the ones included in our study (1 to 2 μg/L).

Despite the advantages of using a comprehensive PBBK model, coupled with exposure reconstruction algorithms to properly interpret the biomonitoring spot urine samples, additional steps are needed towards a comprehensive understanding of the potential adverse effects of BPA. Firstly, we need to always keep in mind that BPA is one of the most rapidly metabolized industrial compounds; thus, to better identify the potential exposure sources, more than one measurements in urine samples per day are required. These are essential for a more accurate description of the time-dependent profile of internal dose and the respective AUC calculation. Moreover, it would be of particular importance to understand how the different exposure regimes and internal exposure profiles induce early biological responses; metabolomics analysis of the respective urine samples and the subsequent pathway analysis using big data analytics and advanced bioinformatics [66] could shed additional light in the potential homeostasis perturbation in relation to BPA exposure. This is the direction of our future work, where the various daily intake estimates will be further associated with responses of molecular targets. Overall, answering the question of whether BPA at current exposure levels eventually results in adverse health outcomes, requires the integration of environmental and molecular epidemiology and toxicology, with detailed exposure-related data (e.g., food questionnaires), multiple daily urine samples collection, internal dose analysis, -omics pathway analysis, and the evaluation of both in vitro and in vivo data. In this content, the capability offered by the PBBK models to estimate BED is expected to provide more accurate exposure metrics for environment-wide association studies, as well as for linking real-life exposure with biologically effective doses in in vitro models [6].

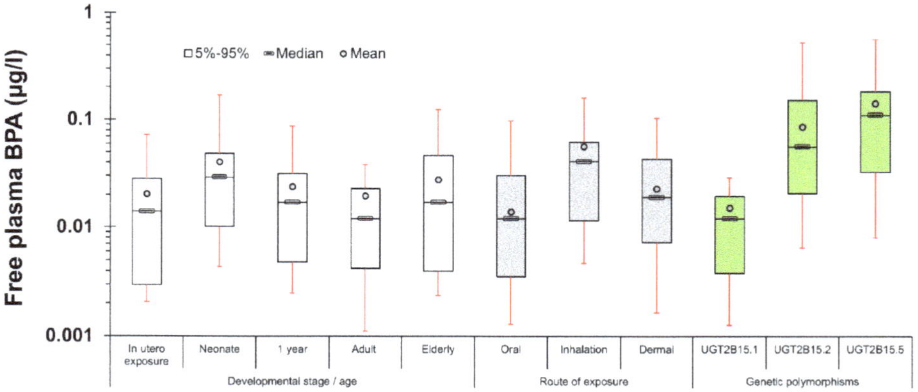

Figure 5. Understanding the major parameters inducing bioavailability differences.

5. Conclusions

Starting from HBM data collection and using a human physiology-based biokinetic model coupled with an exposure reconstruction algorithm, exposure to BPA in the European population was estimated to be almost two orders of magnitude lower than the respective EFSA t-TDI of 4 μg/kg_bw/day. Among the investigated population groups, higher mean intake levels were estimated for children. On the other hand, the similar maximum exposure estimates (close to 0.8 μg/kg_bw/day) for all age groups, indicates that significant exposure sources still occur for both adults and children. On the other hand, these exposure levels might result in a daily AUC that is similar to the one derived by t-TDI, depending on the presence of alleles associated with a slower metabolism of the individuals. Our forthcoming work will focus on BPA substitutes, BPS and BPF, the toxicities of which are yet to be determined, due to uncertainties with their biokinetic and biochemical behaviors especially when treating interactions with specific protein receptors associated with endocrine disruption and the impairment of reproductive capacity.

The multi-faceted method presented herein, provide more accurate daily intake estimates compared to the method based on urinary concentrations mass balance, since the use of physiology-based biokinetic modeling allows the comprehensive description of exposure and elimination time dynamics. This highlights the importance for using PBBK modelling-based exposure reconstruction schemes for rapidly metabolized and excreted compounds, such as BPA. In addition, translating biomonitoring data into internal dose estimates and accounting for the parameters that induce interindividual variability, will provide more accurate exposure metrics for future associations towards a better assessment of potential adverse health effects.

Supplementary Materials: The following are available online at http://www.mdpi.com/2311-5521/4/1/4/s1, Table S1: Regression coefficients for lifetime scaling (from conception to adulthood), Table S2: Gestation parameters (from conception to birth), Table S3: Partition coefficients (organ/blood), Table S4: Model parameters for MCMC analysis.

Author Contributions: D.S. was responsible for the main concept development; the methodology presented, its software implementation, and validation and formal analysis of the results were jointly developed by D.S. and S.K.; the manuscript was jointly prepared by D.S. and S.K.; D.S. further reviewed and edited the final version before submission; finally, S.K. was responsible for all visualizations included in the text.

Funding: This research was funded by the LIFE + Program, grant number LIFE12 ENV/GR/001040 (CROME), and by the 7th RTD Framework Programme, grant agreement No: 603946 (HEALS) and CEFIC LRI, through the DOREMI project.

Conflicts of Interest: The authors declare no conflict of interest.

References

1. Sarigiannis, D.A.; Karakitsios, S.P. A dynamic physiology based pharmacokinetic model for assessing lifelong internal dose. In Proceedings of the AIChE 2012, Pittsburgh, PA, USA, 28 October–2 November 2012.
2. Sarigiannis, D.A.; Gotti, A. Biology-based dose-response models for health risk assessment of chemical mixtures. *Fres. Environ. Bull.* **2008**, *17*, 1439–1451.
3. Georgopoulos, P.G.; Sasso, A.F.; Isukapalli, S.S.; Lioy, P.J.; Vallero, D.A.; Okino, M.; Reiter, L. Reconstructing population exposures to environmental chemicals from biomarkers: Challenges and opportunities. *J. Expo. Sci. Envion. Epidemiol.* **2008**, *19*, 149–171. [CrossRef] [PubMed]
4. Andra, S.S.; Charisiadis, P.; Karakitsios, S.; Sarigiannis, D.A.; Makris, K.C. Passive exposures of children to volatile trihalomethanes during domestic cleaning activities of their parents. *Environ. Res.* **2015**, *136*, 187–195. [CrossRef] [PubMed]
5. Tan, Y.M.; Liao, K.; Conolly, R.; Blount, B.; Mason, A.; Clewell, H. Use of a physiologically based pharmacokinetic model to identify exposures consistent with human biomonitoring data for chloroform. *J. Toxicol. Environ. Health Part A: Curr. Issues* **2006**, *69*, 1727–1756. [CrossRef] [PubMed]
6. Judson, R.S.; Kavlock, R.J.; Setzer, R.W.; Cohen Hubal, E.A.; Martin, M.T.; Knudsen, T.B.; Houck, K.A.; Thomas, R.S.; Wetmore, B.A.; Dix, D.J. Estimating toxicity-related biological pathway altering doses for high-throughput chemical risk assessment. *Chem. Res. Toxicol.* **2011**, *24*, 451–462. [CrossRef] [PubMed]
7. Eissing, T.; Kuepfer, L.; Becker, C.; Block, M.; Coboeken, K.; Gaub, T.; Goerlitz, L.; Jaeger, J.; Loosen, R.; Ludewig, B.; et al. A computational systems biology software platform for multiscale modeling and simulation: Integrating whole-body physiology, disease biology, and molecular reaction networks. *Front. Phys.* **2011**, *2*, 1–10. [CrossRef] [PubMed]
8. Krauss, M.; Schaller, S.; Borchers, S.; Findeisen, R.; Lippert, J.; Kuepfer, L. Integrating Cellular Metabolism into a Multiscale Whole-Body Model. *PLoS Comp. Biol.* **2012**, *8*, e1002750. [CrossRef]
9. Morck, T. Chapter 3G Bisphenol A. In *Biomarkers and Human Biomonitoring*; The Royal Society of Chemistry: London, UK, 2012; Volume 1, pp. 360–380.
10. Rochester, J.R. Bisphenol A and human health: A review of the literature. *Reprod. Toxicol.* **2013**, *42*, 132–155. [CrossRef]

11. Tyl, R.W.; Myers, C.B.; Marr, M.C.; Thomas, B.F.; Keimowitz, A.R.; Brine, D.R.; Veselica, M.M.; Fail, P.A.; Chang, T.Y.; Seely, J.C.; et al. Three-generation reproductive toxicity study of dietary bisphenol A in CD Sprague-Dawley rats. *Toxicol. Sci.* **2002**, *68*, 121–146. [CrossRef]
12. Tyl, R.W.; Myers, C.B.; Marr, M.C.; Sloan, C.S.; Castillo, N.P.; Veselica, M.M.; Seely, J.C.; Dimond, S.S.; Van Miller, J.P.; Shiotsuka, R.N.; et al. Two-generation reproductive toxicity study of dietary bisphenol A in CD-1 (swiss) mice. *Toxicol. Sci.* **2008**, *104*, 362–384. [CrossRef]
13. Ferguson, S.A.; Law, C.D., Jr.; Abshire, J.S. Developmental treatment with bisphenol A or ethinyl estradiol causes few alterations on early preweaning measures. *Toxicol. Sci.* **2011**, *124*, 149–160. [CrossRef] [PubMed]
14. Delclos, K.B.; Camacho, L.; Lewis, S.M.; Vanlandingham, M.M.; Latendresse, J.R.; Olson, G.R.; Davis, K.J.; Patton, R.E.; Gamboa da Costa, G.; Woodling, K.A.; et al. Toxicity evaluation of bisphenol A administered by gavage to Sprague Dawley rats from gestation day 6 through postnatal day 90. *Toxicol. Sci.* **2014**, *139*, 174–197. [CrossRef]
15. Edginton, A.N.; Ritter, L. Predicting plasma concentrations of bisphenol A in children younger than 2 years of age after typical feeding schedules, using a physiologically based toxicokinetic model. *Environ. Health Perspect.* **2009**, *117*, 645–652. [CrossRef] [PubMed]
16. Ginsberg, G.; Rice, D.C. Does rapid metabolism ensure negligible risk from bisphenol A? *Environ. Health Perspect.* **2009**, *117*, 1639–1643. [CrossRef] [PubMed]
17. Sarigiannis, D.; Karakitsios, S.; Gotti, A.; Loizou, G.; Cherrie, J.; Smolders, R.; De Brouwere, K.; Galea, K.; Jones, K.; Handakas, E.; et al. Integra: From global scale contamination to tissue dose. In Proceedings of the 7th International Congress on Environmental Modelling and Software: Bold Visions for Environmental Modeling, San Diego, CA, USA, 15–19 June 2014; pp. 1001–1008.
18. Sarigiannis, D.A.; Papadaki, K.; Kontoroupis, P.; Karakitsios, S.P. Development of QSARs for parameterizing Physiology Based ToxicoKinetic models. *Food Chem. Toxicol.* **2017**, *106*, 114–124. [CrossRef]
19. Papadaki, K.C.; Karakitsios, S.P.; Sarigiannis, D.A. Modeling of adipose/blood partition coefficient for environmental chemicals. *Food Chem. Toxicol.* **2017**, *110*, 274–285. [CrossRef]
20. Beaudouin, R.; Micallef, S.; Brochot, C. A stochastic whole-body physiologically based pharmacokinetic model to assess the impact of inter-individual variability on tissue dosimetry over the human lifespan. *Regul. Toxicol. Pharmacol.* **2010**, *57*, 103–116. [CrossRef]
21. Lee, S.K.; Ou, Y.C.; Andersen, M.E.; Yang, R.S.H. A physiologically based pharmacokinetic model for lactational transfer of PCB 153 with or without PCB 126 in mice. *Arch. Toxicol.* **2007**, *81*, 101–111. [CrossRef]
22. Verner, M.A.; Charbonneau, M.; Lopez-Carrillo, L.; Haddad, S. Physiologically based pharmacokinetic modeling of persistent organic pollutants for lifetime exposure assessment: A new tool in breast cancer epidemiologic studies. *Environ. Health Perspect.* **2008**, *116*, 886–892. [CrossRef]
23. Touitou, E. Drug delivery across the skin. *Expert Opin. Biol. Ther.* **2002**, *2*, 723–733. [CrossRef]
24. Mitragotri, S.; Anissimov, Y.G.; Bunge, A.L.; Frasch, H.F.; Guy, R.H.; Hadgraft, J.; Kasting, G.B.; Lane, M.E.; Roberts, M.S. Mathematical models of skin permeability: An overview. *Int. J. Pharm.* **2011**, *418*, 115–129. [CrossRef] [PubMed]
25. Sarigiannis, D.; Karakitsios, S.; Handakas, E.; Simou, K.; Solomou, E.; Gotti, A. Integrated exposure and risk characterization of bisphenol-A in Europe. *Food Chem. Toxicol.* **2016**, *98*, 134–147. [CrossRef] [PubMed]
26. Edginton, A.N.; Schmitt, W.; Voith, B.; Willmann, S. A mechanistic approach for the scaling of clearance in children. *Clin. Pharmacokinet.* **2006**, *45*, 683–704. [CrossRef] [PubMed]
27. Leeder, J.S. Developmental pharmacogenetics: A general paradigm for application to neonatal pharmacology and toxicology. *Clin. Pharmacol. Ther.* **2009**, *86*, 678–682. [CrossRef]
28. Court, M.H.; Zhang, X.; Ding, X.; Yee, K.K.; Hesse, L.M.; Finel, M. Quantitative distribution of mRNAs encoding the 19 human UDP-glucuronosyltransferase enzymes in 26 adult and 3 fetal tissues. *Xenobiotica* **2012**, *42*, 266–277. [CrossRef] [PubMed]
29. Völkel, W.; Colnot, T.; Csanády, G.A.; Filser, J.G.; Dekant, W. Metabolism and kinetics of bisphenol a in humans at low doses following oral administration. *Chem. Res. Toxicol.* **2002**, *15*, 1281–1287.
30. Thayer, K.A.; Doerge, D.R.; Hunt, D.; Schurman, S.H.; Twaddle, N.C.; Churchwell, M.I.; Garantziotis, S.; Kissling, G.E.; Easterling, M.R.; Bucher, J.R.; et al. Pharmacokinetics of bisphenol A in humans following a single oral administration. *Environ. Int.* **2015**, *83*, 107–115. [CrossRef]

31. Teeguarden, J.G.; Twaddle, N.C.; Churchwell, M.I.; Yang, X.; Fisher, J.W.; Seryak, L.M.; Doerge, D.R. 24-hour human urine and serum profiles of bisphenol A: Evidence against sublingual absorption following ingestion in soup. *Toxicol. Appl. Pharmacol.* **2015**, *288*, 131–142. [CrossRef]
32. Yang, X.; Doerge, D.R.; Teeguarden, J.G.; Fisher, J.W. Development of a physiologically based pharmacokinetic model for assessment of human exposure to bisphenol A. *Toxicol. Appl. Pharmacol.* **2015**, *289*, 442–456. [CrossRef]
33. International Commission on Radiological Protection (ICRP). *Basic Anatomical and Physiological Data for Use in Radiological Protection: Reference Values*; ICRP: Ottawa, ON, Canada, 2002.
34. Sarigiannis, D.; Karakitsios, S.; Gotti, A.; Handakas, E. Life cycle-based health risk assessment of plastic waste. In Proceedings of the 5th International Conference on Sustainable Solid Waste Manage, Athens, Greece, 21–24 June 2017.
35. Gilks, W.R.; Roberts, G.O. Strategies for improving MCMC. In *Markov Chain Monte Carlo in Practice*; Springer: Berlin, Germany, 1996; pp. 89–114. [CrossRef]
36. Haario, H.; Laine, M.; Mira, A.; Saksman, E. DRAM: Efficient adaptive MCMC. *Stat. Comp.* **2006**, *16*, 339–354. [CrossRef]
37. Ter Braak, C.J. A Markov Chain Monte Carlo version of the genetic algorithm Differential Evolution: Easy Bayesian computing for real parameter spaces. *Stat. Comp.* **2006**, *16*, 239–249. [CrossRef]
38. Vandentorren, S.; Zeman, F.; Morin, L.; Sarter, H.; Bidondo, M.L.; Oleko, A.; Leridon, H. Bisphenol-A and phthalates contamination of urine samples by catheters in the Elfe pilot study: Implications for large-scale biomonitoring studies. *Environ. Res.* **2011**, *111*, 761–764. [CrossRef] [PubMed]
39. Federal Environment Agency (UBA). The German Environment Specimen Bank. Available online: http://www.umweltprobenbank.de (accessed on 17 April 2017).
40. DEMOCOPHES. DEMOCOPHES Layman's Report—Human Biomonitoring on a European Scale. Available online: http://www.eu-hbm.info/euresult/layman-report (accessed on 17 April 2017).
41. Covaci, A.; Hond, E.D.; Geens, T.; Govarts, E.; Koppen, G.; Frederiksen, H.; Knudsen, L.E.; Mørck, T.A.; Gutleb, A.C.; Guignard, C.; et al. Urinary BPA measurements in children and mothers from six European member states: Overall results and determinants of exposure. *Environ. Res.* **2015**, *141*, 77–85. [CrossRef] [PubMed]
42. Frederiksen, H.; Jensen, T.K.; Jørgensen, N.; Kyhl, H.B.; Husby, S.; Skakkebæk, N.E.; Main, K.M.; Juul, A.; Andersson, A. Human urinary excretion of non-persistent environmental chemicals: An overview of Danish data collected between 2006 and 2012. *Reproduction* **2014**, *147*, 555–565. [CrossRef]
43. Becker, K.; Göen, T.; Seiwert, M.; Conrad, A.; Pick-Fuß, H.; Müller, J.; Wittassek, M.; Schulz, C.; Kolossa-Gehring, M. GerES IV: Phthalate metabolites and bisphenol A in urine of German children. *Int. J. Hyg. Environ. Health* **2009**, *212*, 685–692. [CrossRef] [PubMed]
44. Galloway, T.; Cipelli, R.; Guralnik, J.; Ferrucci, L.; Bandinelli, S.; Corsi, A.M.; Money, C.; McCormack, P.; Melzer, D. Daily bisphenol a excretion and associations with sex hormone concentrations: Results from the InCHIANTI adult population study. *Environ. Health Perspect.* **2010**, *118*, 1603–1608. [CrossRef] [PubMed]
45. Ye, X.; Pierik, F.H.; Hauser, R.; Duty, S.; Angerer, J.; Park, M.M.; Burdorf, A.; Hofman, A.; Jaddoe, V.W.V.; Mackenbach, J.P.; et al. Urinary metabolite concentrations of organophosphorous pesticides, bisphenol A, and phthalates among pregnant women in Rotterdam, the Netherlands: The Generation R study. *Environ. Res.* **2008**, *108*, 260–267. [CrossRef] [PubMed]
46. Casas, L.; Fernandez, M.F.; Llop, S.; Guxens, M.; Ballester, F.; Olea, N.; Irurzun, M.B.; Rodriguez, L.S.; Riano, I.; Tardon, A.; et al. Urinary concentrations of phthalates and phenols in a population of Spanish pregnant women and children. *Environ. Int.* **2011**, *37*, 858–866. [CrossRef]
47. Dereumeaux, C.; Fillol, C.; Charles, M.-A.; Denys, S. The French human biomonitoring program: First lessons from the perinatal component and future needs. *Int. J. Hyg. Environ. Health* **2017**, *220*, 64–70. [CrossRef]
48. Myridakis, A.; Fthenou, E.; Balaska, E.; Vakinti, M.; Kogevinas, M.; Stephanou, E.G. Phthalate esters, parabens and bisphenol-A exposure among mothers and their children in Greece (Rhea cohort). *Environ. Int.* **2015**, *83*, 1–10. [CrossRef]
49. EFSA. Scientific Opinion on the risks to public health related to the presence of bisphenol A (BPA) in foodstuffs. *EFSA J.* **2015**, *13*, 3978. [CrossRef]
50. Krishnan, K.; Gagné, M.; Nong, A.; Aylward, L.L.; Hays, S.M. Biomonitoring Equivalents for bisphenol A (BPA). *Regul. Toxicol. Pharmacol.* **2010**, *58*, 18–24. [CrossRef] [PubMed]

51. Hays, S.M.; Becker, R.A.; Leung, H.W.; Aylward, L.L.; Pyatt, D.W. Biomonitoring equivalents: A screening approach for interpreting biomonitoring results from a public health risk perspective. *Regul. Toxicol. Pharmacol.* **2007**, *47*, 96–109. [CrossRef] [PubMed]
52. Aylward, L.L. Integration of biomonitoring data into risk assessment. *Curr. Opin. Toxicol.* **2018**, *9*, 14–20. [CrossRef]
53. Geens, T.; Apelbaum, T.Z.; Goeyens, L.; Neels, H.; Covaci, A. Intake of bisphenol A from canned beverages and foods on the Belgian market. *Food Addit. Contam.: Part A* **2010**, *27*, 1627–1637. [CrossRef]
54. Sakhi, A.K.; Lillegaard, I.T.; Voorspoels, S.; Carlsen, M.H.; Loken, E.B.; Brantsaeter, A.L.; Haugen, M.; Meltzer, H.M.; Thomsen, C. Concentrations of phthalates and bisphenol A in Norwegian foods and beverages and estimated dietary exposure in adults. *Environ. Int.* **2014**, *73*, 259–269. [CrossRef]
55. von Goetz, N.; Wormuth, M.; Scheringer, M.; Hungerbuhler, K. Bisphenol A: How the most relevant exposure sources contribute to total consumer exposure. *Risk Anal. Int. J.* **2010**, *30*, 473–487. [CrossRef]
56. Cao, X.L.; Perez-Locas, C.; Dufresne, G.; Clement, G.; Popovic, S.; Beraldin, F.; Dabeka, R.W.; Feeley, M. Concentrations of bisphenol A in the composite food samples from the 2008 Canadian total diet study in Quebec City and dietary intake estimates. *Food Addit. Contam.* **2011**, *28*, 791–798. [CrossRef]
57. Lorber, M.; Schecter, A.; Paepke, O.; Shropshire, W.; Christensen, K.; Birnbaum, L. Exposure assessment of adult intake of bisphenol A (BPA) with emphasis on canned food dietary exposures. *Environ. Int.* **2015**, *77*, 55–62. [CrossRef]
58. LaKind, J.S.; Naiman, D.Q. Bisphenol A (BPA) daily intakes in the United States: Estimates from the 2003–2004 NHANES urinary BPA data. *J. Expo. Sci. Environ. Epidemiol.* **2008**, *18*, 608–615. [CrossRef]
59. LaKind, J.S.; Naiman, D.Q. Temporal trends in bisphenol A exposure in the United States from 2003–2012 and factors associated with BPA exposure: Spot samples and urine dilution complicate data interpretation. *Environ. Res.* **2015**, *142*, 84–95. [CrossRef] [PubMed]
60. Fisher, J.W.; Twaddle, N.C.; Vanlandingham, M.; Doerge, D.R. Pharmacokinetic modeling: Prediction and evaluation of route dependent dosimetry of bisphenol A in monkeys with extrapolation to humans. *Toxicol. Appl. Pharmacol.* **2011**, *257*, 122–136. [CrossRef] [PubMed]
61. Hanioka, N.; Oka, H.; Nagaoka, K.; Ikushiro, S.; Narimatsu, S. Effect of UDP-glucuronosyltransferase 2B15 polymorphism on bisphenol A glucuronidation. *Arch. Toxicol.* **2011**, *85*, 1373–1381. [CrossRef] [PubMed]
62. Partosch, F.; Mielke, H.; Gundert-Remy, U. Functional UDP-glucuronyltransferase 2B15 polymorphism and bisphenol A concentrations in blood: Results from physiologically based kinetic modelling. *Arch. Toxicol.* **2013**, *87*, 1–8. [CrossRef] [PubMed]
63. Teeguarden, J.; Hanson-Drury, S.; Fisher, J.W.; Doerge, D.R. Are typical human serum BPA concentrations measurable and sufficient to be estrogenic in the general population? *Food Chem. Toxicol.* **2013**, *62*, 949–963. [CrossRef] [PubMed]
64. Teeguarden, J.G.; Calafat, A.M.; Ye, X.; Doerge, D.R.; Churchwell, M.I.; Gunawan, R.; Graham, M.K. Twenty-four hour human urine and serum profiles of bisphenol a during high-dietary exposure. *Toxicol. Sci.* **2011**, *123*, 48–57. [CrossRef] [PubMed]
65. Teeguarden, J.G.; Twaddle, N.C.; Churchwell, M.I.; Doerge, D.R. Urine and serum biomonitoring of exposure to environmental estrogens I: Bisphenol A in pregnant women. *Food Chem. Toxicol.* **2016**, *92*, 129–142. [CrossRef]
66. Manrai, A.K.; Cui, Y.; Bushel, P.R.; Hall, M.; Karakitsios, S.; Mattingly, C.; Ritchie, M.; Schmitt, C.; Sarigiannis, D.A.; Thomas, D.C.; et al. Informatics and Data Analytics to Support Exposome-Based Discovery for Public Health. *Ann. Rev. Public Health* **2016**. [CrossRef]

fluids

MDPI

Article

Computational Study of the Interaction of a PEGylated Hyperbranched Polymer/Doxorubicin Complex with a Bilipid Membrane

Prodromos Arsenidis and Kostas Karatasos *

Physical Chemistry Laboratory, Chemical Engineering Department, Aristotle University of Thessaloniki, 54124 Thessaloniki, Greece; arsenidisp@gmail.com

* Correspondence: karatas@eng.auth.gr; Tel.: +30-231-099-5850

Received: 13 December 2018; Accepted: 19 January 2019; Published: 24 January 2019

Abstract: Fully atomistic molecular dynamics simulations are employed to study in detail the interactions between a complex comprised by a PEGylated hyperbranched polyester (HBP) and doxorubicin molecules, with a model dipalmitoylphosphatidylglycerol membrane in an aqueous environment. The effects of the presence of the lipid membrane in the drug molecules' spatial arrangement were examined in detail and the nature of their interaction with the latter were discussed and quantified where possible. It was found that a partial migration of the drug molecules towards the membrane's surface takes place, driven either by hydrogen-bonding (for the protonated drugs) or by hydrophobic interactions (for the neutral drug molecules). The clustering behavior of the drug molecules appeared to be enhanced in the presence of the membrane, while the development of a charge excess close to the surface of the hyperbranched polymer and of the lipid membrane was observed. The uneven charge distribution created an effective overcharging of the HBP/drug complex and the membrane/drug surface. The translational motion of the drug molecules was found to be strongly affected by the presence of the membrane. The extent of the observed changes depended on the charge of the drug molecule. The build-up of the observed charge excesses close to the surface of the polymeric host and the membrane, together with the changes in the diffusional behavior of the drug molecules are of particular interest. Both phenomena could be important at the latest stages of the liposomal disruption and the release of the drug cargo in formulations based on relevant liposomal carriers.

Keywords: dipalmitoylphosphatidylglycerol (DPPG); doxorubicin; hyperbranched polyester; simulations

1. Introduction

Vesicles based on lipids are recognized as promising non-viral vectors for drug or gene delivery purposes [1–5]. In particular, phospholipid liposomes are considered as suitable vehicles for pharmaceutical compounds due to their biodegradable and biocompatible nature, which is combined with low toxicity levels [6]. In such liposome/drug formulations, the nature of the interactions between the guest molecules and the lipid vesicle is of paramount significance since it is intimately related to the ability of the carrier to minimize the loss of the bioactive cargo during the transfer to the desired destination, as well as to the mechanism of its liposomal escape at the target site [7,8].

Molecular simulations have started playing an increasingly important role in the description of the interactions of bioactive molecules with lipid layers [9–12], due to their ability to provide information at the molecular or even at the atomic level. Among other model phospholipid layers, dipalmitoylphosphatidylglycerol (DPPG)-based systems have been recently studied in drug/liposome formulations [13]. DPPG membranes (formed by a double layer of lipids, i.e., a bilipid layer) carry a negative charge per lipid at physiological pH conditions [14,15] which renders them good candidates when complexation with cationic bioactive compounds is desired [13,14,16]. This attribute of DPPG lipids can be exploited in more complex drug/liposome formulations, where optimization of the drug loading conditions is promoted by the presence of multifunctional molecules such as hyperbranched polymers [4,17,18].

In this work we examine by means of fully atomistic molecular dynamics simulations, the interactions of a complex comprised by a functionalized (PEGylated) hyperbranched polymer (HBP) and an anticancer drug, doxorubicin, with a model bilipid DPPG membrane in an aqueous environment. Structural and dynamic characteristics of the hyperbranched-polymer/doxorubicin aqueous solution have been investigated in our previous work [19]. We henceforth aim at elucidating the effects of the presence of the DPPG bilayer to the drug spatial distribution with respect to the polymer and the lipid membrane, the nature of the drug/DPPG interactions, the effective charge distributions related to possible overcharging phenomena of the polymer and of the bilipid DPPG layer, and the changes in the translational behavior of the drug molecules imparted by the presence of the bilipid membrane.

2. Materials and Methods

Initial configuration of the DPPG bilipid membrane was taken from a pre-equilibrated system of a several hundred ns long simulation, from a publicly available repository [20]. Pre-equilibration of the hydrated DPPG membranes were performed by Jämbeck and collaborators [21–23] in explicit TIP3P [24] water solution. Figure 1 illustrates a single DPPG lipid, a doxorubicin molecule, and the structure of the PEGylated hyperbranched polymer.

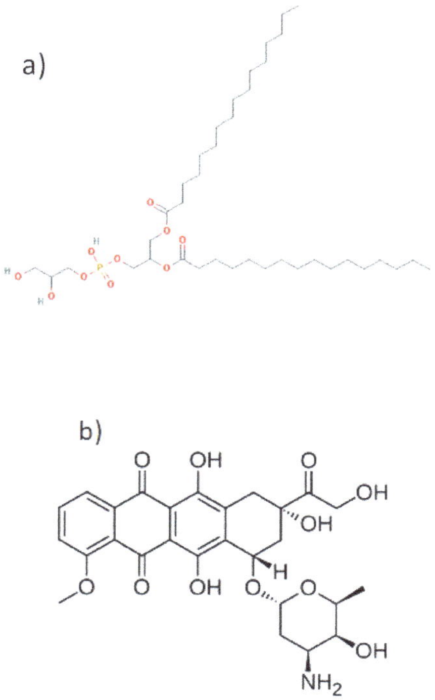

Figure 1. *Cont.*

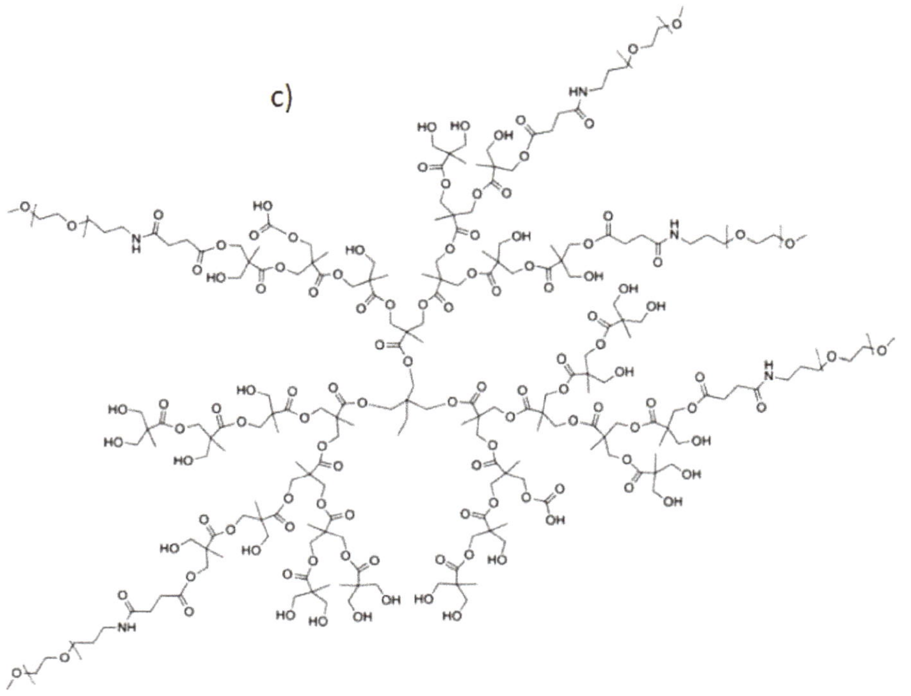

Figure 1. (**a**) A single dipalmitoylphosphatidylglycerol (DPPG) lipid. (**b**) Doxorubicin. (**c**) The structure of the PEGylated hyperbranched polyester. Each PEG arm consisted of 20 monomers.

Figure 2 depicts the equilibrated DPPG membrane comprised by 128 lipids. As a next step, the bilipid membrane was combined with the HBP/Doxorubicin complex that was taken from our previous study [19] with the aid of the VEGA ZZ software [25]. The HBP/Doxorubicin complex was initially placed on top of one side of the DPPG bilayer with a separation of approximately 65 Å between the center of mass of the HBP and that of the membrane (no constraints have been imposed to any of the constituents during the equilibration or the production runs). For comparison purposes, we have kept exactly the same number of neutral and protonated drug molecules and the same ionic strength of 165 mM as in Reference [19]. Figure 3 depicts the initial configuration of the bilipid/hyperbranched/drugs/counterions system, with (**a**) and without (**b**) the water molecules. The number of the different components comprising the examined system are presented in Table 1.

Table 1. Number of molecules/ions of each type included in the simulated system.

PEGylated Hyperbranched Polymer	DPPG Lipids	Neutral Drug Molecules	Positively Charged Drugs	Water	Chlorine Ions	Sodium Ions
1	128	19	31	54234	35	132

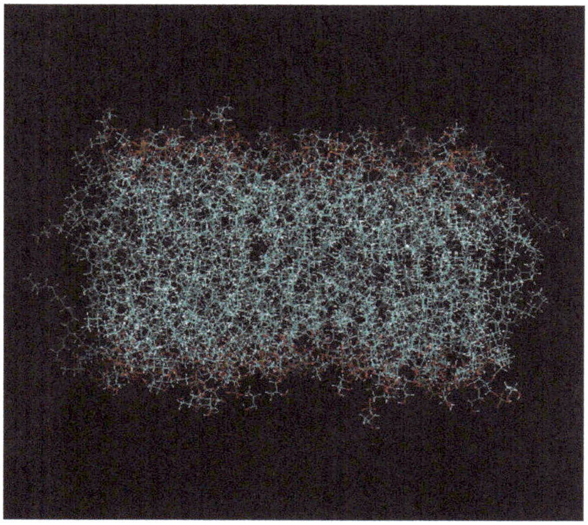

Figure 2. The pre-equilibrated model of the DPPG membrane that was used in our study.

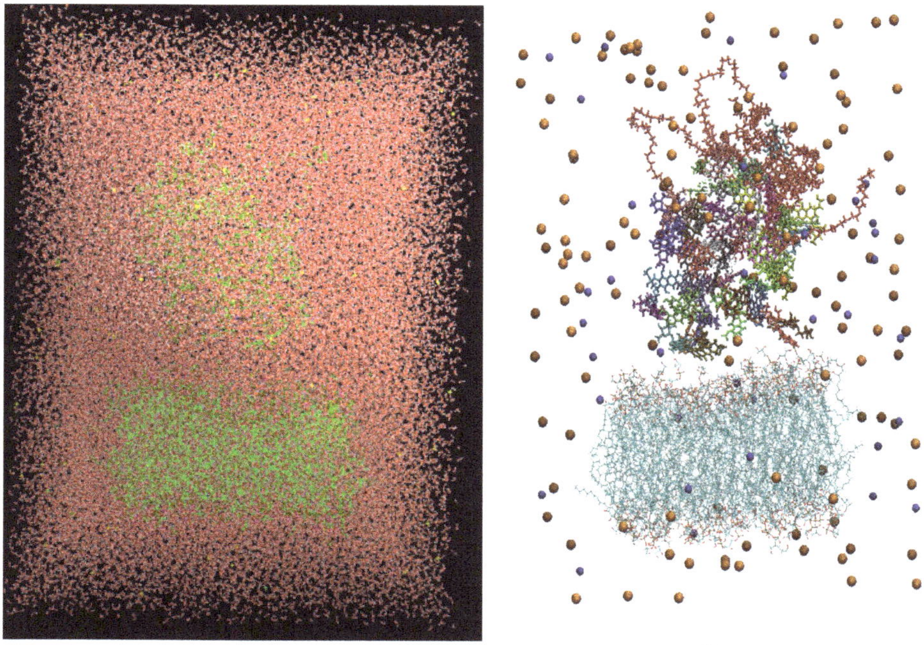

(a) (b)

Figure 3. (**a**) The constructed system including water molecules (shown in red-white). (**b**) The same system without the water molecules. Blue beads represent the chlorine ions and orange beads the sodium ions. Each drug molecule is shown in different color and in rod representation, while the PEGylated hyperbranched molecule is shown in red color. In the lower part of the snapshot, the lipids of the membrane are shown in wire representation.

The forcefield parameters for the hyperbranched polymer and the drug molecules were taken from the generalized Amber forcefield (GAFF) [26] and for the water molecules from the TIP3P model [24] as described in Reference [19]. For the bilipid membrane, the GAFF lipid set of parameters was utilized, which was developed specifically for the description of lipids [27]. All simulations were performed by the aid of the NAMD simulation package [28] with a timestep of 1 fs and a saving frequency of 2 ps, using periodic boundary conditions. Temperature control was performed using the Langevin method (with a damping coefficient of 5 ps^{-1}) [29]. Electrostatic interactions were calculated via the particle mesh Ewald (PME) scheme [30], while the distance cutoff for the van der Waals interactions was set at 12 Å. The percentage of the protonated doxorubicin molecules was calculated through the Henderson−Hasselbalch relation [31] taking into account the pK_a of the drug's ionizable group at 37 °C and considering physiological pH conditions. An appropriate amount of chlorine and sodium ions was added for the overall neutralization of the system and for accomplishing an ionic strength of 165 mM [32].

More detailed information regarding the construction and the equilibration of the hyperbranched polymer/drug complex can be found in Reference [19]. Following the construction of the system, energy minimization with steepest descent and conjugate gradient methods was applied, prior to 40 ns long MD equilibration runs in the canonical (NVT) ensemble at T = 310 K. This procedure ensured stabilization of energetic (total energy), thermodynamic (pressure), and conformational properties (i.e., average size of the PEGylated hyperbranched polymer) of the polymeric component of the system. The equilibrium area per lipid was approximately 75 $Å^2$ and the parallel orientation of the lipids was found to persist throughout the simulation. Production runs were also performed in the NVT ensemble and at the same (T = 310 K) temperature.

3. Results

The analysis performed in the simulation data aimed at investigating the spatial arrangement of the drug molecules with respect to the hyperbranched carrier and the negatively charged DPPG membrane, the degree of their association with the latter, and the driving forces responsible for the observed association.

3.1. Spatial Distribution of Drug Molecules

To examine the spatial arrangement of the drug molecules with respect to the hyperbranched nanoparticle and the bilipid membrane, we have calculated relevant average distances between centers of mass as well as relevant mass distributions. Figure 4 shows the average distance between the center of mass of the DPPG membrane and the center of mass of the other moieties.

If we take into account that the boundary of the HBP periphery was found to be at approximately 20 Å distance from its center of mass, it appears that drug molecules remain on average close to the HBP's surface. However, the protonated doxorubicin molecules together with the Na^+ counterions appear to be somewhat closer to the membrane. The closer proximity of the positively charged moieties should be attributed to the favorable Coulombic interactions with the membrane. Lack of such favorable electrostatic interactions between the DPPG membrane and the HBP, as well as the repulsive interactions with the Cl^- counterions, can account for the larger distances between those moieties and the bilipid membrane.

To elaborate more on the spatial arrangement of the drug molecules with respect to the hyperbranched nanoparticle, we have calculated their mass distribution as a function of their distance from the center of mass of the HBP, as illustrated in Figure 5.

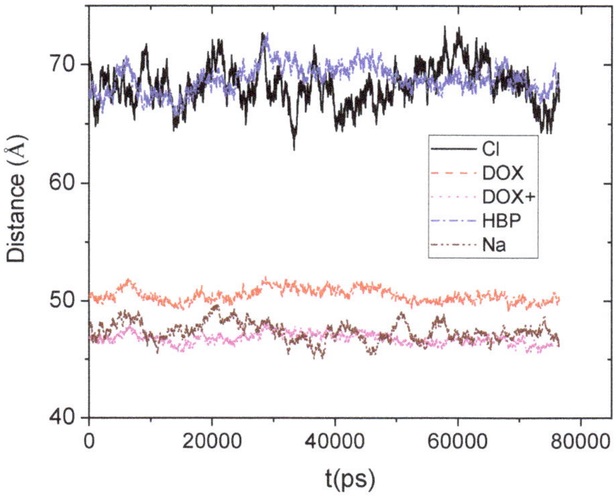

Figure 4. Average distance as a function of time between the center of mass of the lipid membrane (considered as an individual molecule including all the lipids) and those of the hyperbranched polymer, the counterions (Cl^-, Na^+), and the neutral (DOX) and the protonated (DOX+) drug molecules.

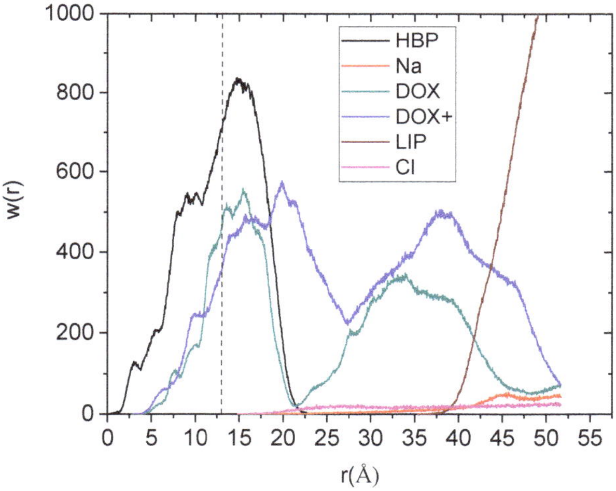

Figure 5. Weight distribution of the drug molecules and of the other moieties with respect to the center of mass of the hyperbranched polymer (a spherical symmetry of the hyperbranched polyester (HBP) molecule has been considered). The notation in the legend is as in Figure 4, with LIP denoting the DPPG membrane. The vertical dashed line denotes the location of the radius of gyration of the polymer.

The main feature characterizing the mass distributions of the drug molecules is their bimodal behavior. No such bimodal behavior for the protonated doxorubicin was noted in the absence of the membrane [19]. The second peak of the drugs' distribution located beyond the HBP boundary exhibits a partial overlap with the membrane distribution. The degree of this overlap is considerably higher between the distribution describing the protonated drug molecules and that of the membrane. Quantification of the degree of penetration of the drug molecules within the HBP's structure and within the bilipid membrane can be made by calculating the area of the profiles within the boundaries of the HBP and the membrane respectively. Tables 2 and 3 list the so-calculated degrees of drug penetration.

Table 2. Average number and degree of penetration of drug molecules within the HBP's structure.

Molecular Species	Number of Molecules	Percentage (%)
DOX	8	42
DOX+	11	35

Table 3. Average number and degree of penetration of drug molecules within the membrane's structure.

Molecular Species	Number of Molecules	Percentage (%)
DOX	3.6	19
DOX+	10	30

The percentages appearing in Table 3 indicate that a significant amount of drug molecules have migrated from the HBP/drug complex towards the DPPG surface. This percentage is considerably higher for the protonated drug molecules, apparently due to the attractive Coulombic interactions with the negatively charged lipids. It must be noted here, that the percentages appearing in Tables 2 and 3 reflect equilibrium conditions, since the timescale of the simulation is orders of magnitude longer compared to the timescale necessary for the drug molecules to reach the diffusive regime at conditions similar to the ones examined here [19], while the average distances between the different molecules remain stable (see Figure 4).

3.2. Charge Distributions

In our effort on one hand to discuss further the driving forces which are responsible for the mass distributions described before, and on the other hand to explore overcharging phenomena which are known to play a significant role in drug or gene-delivery formulations [33–35] and in the conformational properties of the hyperbranched molecules [36], we calculated component and overall charge distributions close to the HBP/drug complex and close to the membrane's surface. Figure 6 portrays the overall charge profile (arising from all components in the system) when moving radially outward with respect to the center of mass of the HBP (all charges henceforth are presented in units of an electron charge). It is shown that the overall charge profile exhibits a strong fluctuation around zero at distances extending from the center of mass of the PEGylated hyperbranched polyester to a region close to its periphery. At the outermost region, however, a net positive charge develops. As it was found, this charge arises from the positively charged drug molecules which are concentrated close to the HBP's outer surface. Calculation of the overall charge of the complex rendered a net charge of approximately 8.5e (integration from the polymer's center of mass to its boundary at ~22.5 Å).

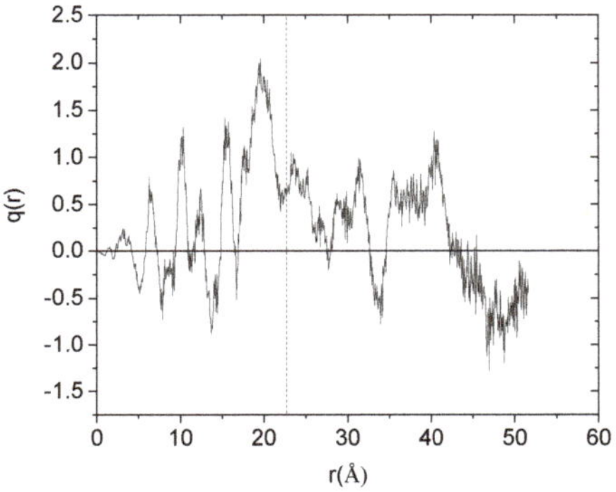

Figure 6. Overall charge distribution with respect to the HBP center of mass arising from all the components in the system. The vertical dashed line denotes the surface boundary of the hyperbranched nanoparticle.

To explore the charge profiles with respect to the bilipid's surface, we considered a reference frame with its origin in the center of mass of the bilayer and with the directions of its axes defined by the principal axes of inertia of the membrane. The profiles, shown in Figure 7, were calculated along the direction normal to the plane defined by the directions of two of the principal axes of inertia which lie parallel to the membrane's surface (here denoted as z).

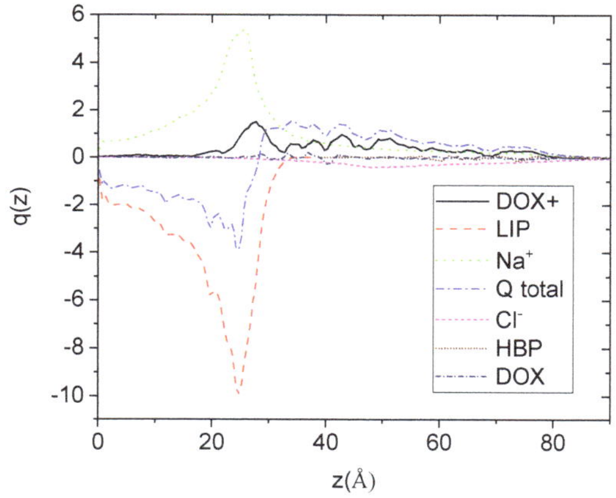

Figure 7. Symmetrized charge distributions of all the moieties in the examined system, as a function of the distance from the plane, which is defined from the center of mass of the lipid membrane (considered as an individual molecule) and the two principal inertia axes, which are parallel to the lipid layer. "Q total" denotes the sum of all the component charge profiles.

Focusing on the overall sum of the distributions, it appears that there exists a net negative charge close to the membrane's surface. The membrane boundary is located close to 30 Å as this has been determined independently by the corresponding mass distribution (not shown here). As anticipated, the net charge arises by the negatively charged lipids. The concentration of positive charges arising from the Na^+ and the DOX+ drug molecules does not suffice to counterbalance the lipids' negative charge. Estimation of the overall charge of the membrane complex results to a value of approximately -100 (integration from the membrane's center of mass to its boundary in both directions), which is appreciably reduced (by about 22%) compared to the entire membrane charge (-128).

3.3. *Effects of the Presence of the Membrane in the Clustering Behavior of Doxorubicin*

As has been demonstrated in past studies (see Reference [19] and references therein) Doxorubicin tends to self-associate in aqueous solutions forming oligomeric clusters. This behavior is attributed to hydrophobic forces and to the $\pi-\pi$ electron interaction between the planar aromatic parts of the drug molecules. The formation of such drug aggregates and the structural characteristics of these clusters may be of paramount importance regarding their pharmacological efficiency, since they can affect their transport properties, their permeation through a membrane, and ultimately the final drug biodistribution. To monitor the changes imparted in the associative behavior of doxorubicin molecules in the presence of the polymeric host and the bilipid membrane, the relative arrangement of the drug molecules was examined through radial distribution functions arising from their centers of mass.

Figure 8 shows the radial distribution functions of the center of mass of the drug molecules (neutral and protonated) with and without the presence of the lipid membrane (the behavior without the presence of the membrane was taken from Reference [19]). The distributions without the membrane are almost bimodal, characterized by a sharp peak close to an intermolecular distance of 5 Å and a second maximum close to 7–8 Å. The short-distance peak arises from the closest neighbors of a drug molecule and is consistent with the formation of dimers, while the peak corresponding to longer distances arises from the presence of neighboring clusters. It must be noted that in the case of the protonated drugs, the electrostatic repulsions due to the presence of the charged amine groups affect the spatial arrangement of the molecules (i.e., the relative intensity and the number of the peaks).

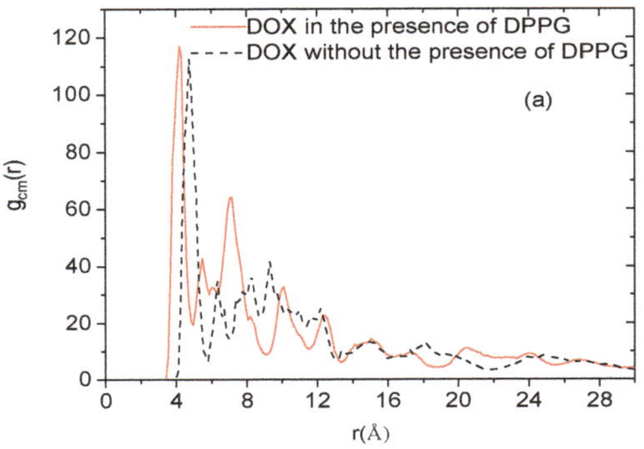

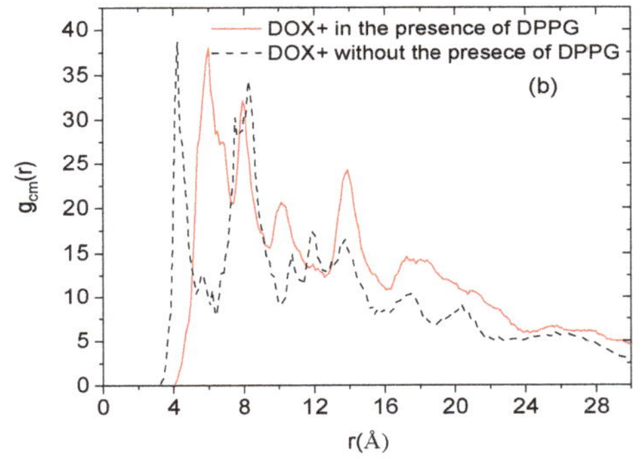

Figure 8. Radial distribution functions of the center of mass of the doxorubicin molecules in the presence (solid lines) and in the absence (dashed lines) of the bilipid membrane, for the (**a**) neutral and the (**b**) protonated drug molecules.

In the case of the protonated drugs in the presence of the membrane, it is observed that on average the separation between the formed clusters decreases (see the distance between the first and the second peak with and without the presence of the membrane). This can be associated with the electrostatic attraction of the protonated drug molecules by the membrane, which results in an enhancement of their tendency to cluster close to the lipid surface. This behavior could also be related to the effective higher concentration of the drug molecules in the system with the membrane. As far as the neutral drugs are concerned, a similar behavior can be observed. It is noticed that in the presence of the DPPG bilayer, the third peak arising from neighboring clusters exhibits a higher intensity compared to the analogous peak without the presence of the membrane. In other words, the limited diffusion of the drug clusters due to the geometric restriction imposed by the membrane, combined with hydrophobic forces, results in a higher degree of the drug molecules' localization. In general, therefore, for both cases (protonated and neutral drugs), the presence of the bilipid membrane appears to reinforce the drug clustering behavior.

3.4. Hydrogen Bonding

In the absence of the DPPG membrane, it was found that apart from the geometric entrapment of the drug molecules within the polymeric structure, another mechanism, that of hydrogen bonding, was also responsible for the association of doxorubicin with the polymeric host [19]. This route for complexation can play a key role in drug delivery processes [37]. In the presence of the bilipid membrane, additional hydrogen-bonding-capable groups are introduced in the system, creating thus antagonistic to the HBP centers for the association of drug molecules. Recalling the image discussed in Section 3.1 (Figure 5 and Table 3), it is of interest to examine the role of hydrogen bonding in the membrane/drug association. To this end, we have examined pair correlation functions of hydrogen-bonding-capable atoms. The criteria we followed for the identification of a hydrogen bond were based on the hydrogen-acceptor distance in combination with the angle formed by the donor-hydrogen-acceptor triplet. The maximum distance considered, was the extent of the corresponding peak in the pertinent pair correlation function, while the minimum donor-hydrogen-acceptor angle was taken to be 120° [38,39].

Such pair correlation functions between atoms belonging to the protonated doxorubicin molecules and to the lipids of the membranes are shown in Figure 9. In all cases, the existence of a sharp peak close to 2.5 Å implies the presence of hydrogen bonds between the examined atomic pairs. The intensity of the hydrogen-bonding peak is indicative to the abundance of each kind of hydrogen bond. Therefore, the pair between the amine hydrogen of the DOX+ and the phosphorus atom of the lipid, appears to be the more abundant. Analogous pair correlation functions between the neutral drug molecules and the lipid are shown in Figure 10.

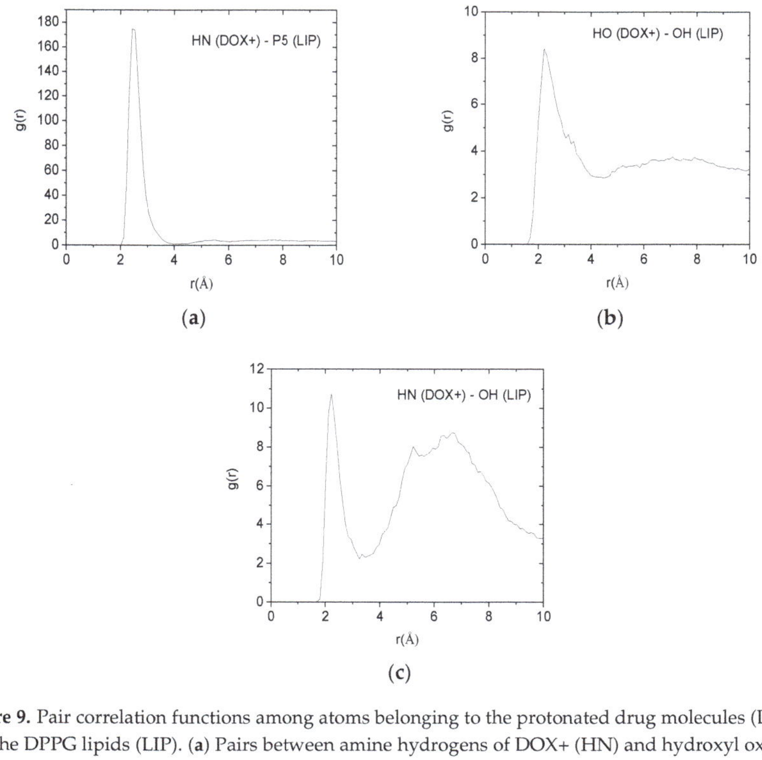

Figure 9. Pair correlation functions among atoms belonging to the protonated drug molecules (DOX+) and the DPPG lipids (LIP). (**a**) Pairs between amine hydrogens of DOX+ (HN) and hydroxyl oxygens of the lipid (OH). (**b**) Pairs between hydroxyl hydrogens (HO) of DOX+ and hydroxyl oxygens (OH) of the lipid. (**c**) Pairs between amine hydrogens of DOX+ (HN) and phosphorus (P5) of the lipid.

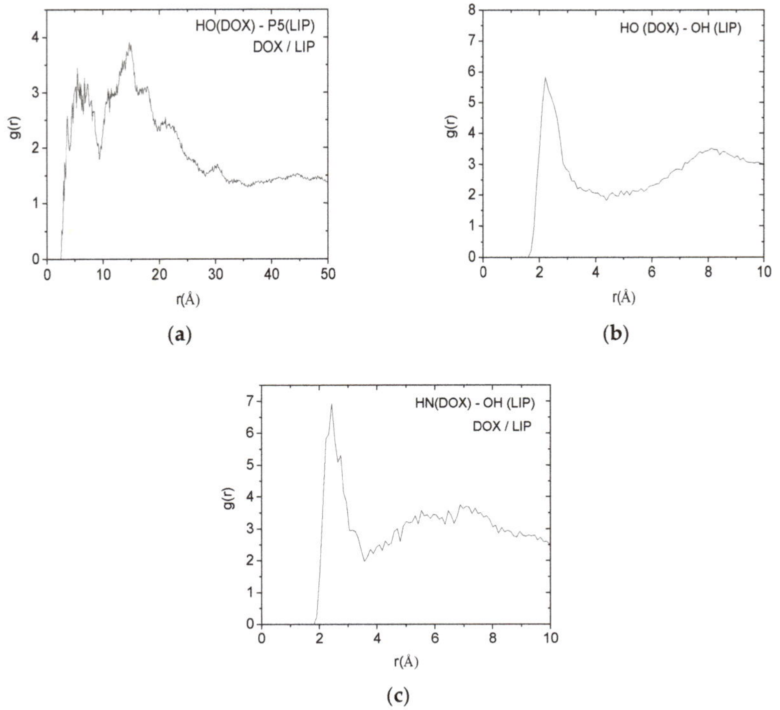

Figure 10. Pair correlation functions among atoms belonging to the neutral drug molecules (DOX) and the DPPG lipids (LIP). (**a**) Pairs between amine hydrogens of DOX (HN) and hydroxyl oxygens of the lipid (OH). (**b**) Pairs between hydroxyl hydrogens (HO) of DOX and hydroxyl oxygens (OH) of the lipid. (**c**) Pairs between amine hydrogens of DOX (HN) and phosphorus (P5) of the lipid.

To quantify the tendency for hydrogen-bond formation between the examined pairs, we have calculated the average number of each kind of hydrogen bond, per timestep, as presented in Table 4.

Table 4. Average number of hydrogen bonds per timestep for the examined pairs.

Drug/LIP	HN (Drug)-OH(LIP)	HN (Drug)-P5(LIP)	HO (Drug)-OH(LIP)
DOX+/LIP	1.6 ± 1.1	10.1 ± 2.2	2.5 ± 1.5
DOX/LIP	0.5 ± 0.8	0.0 ± 0.2	0.8 ± 0.9

Inspection of the values in Table 4 verifies that hydrogen bonding between DOX+ amine hydrogens (HN) and lipid phosphorus atoms (P5) is the most frequent. A lower degree of hydrogen bonding which appears to be marginally statistically significant, is that of the DOX+ hydroxyl hydrogen (HO) and lipid hydroxyl oxygen (OH), while an even lower degree of hydrogen bonding is observed between the DOX+ amine hydrogen (HN) and the lipid's hydroxyl oxygen (OH). In conjunction with the results from Section 3.1 it can be concluded that one of the mechanisms which is responsible for the association between the protonated drug molecules and the membrane's surface is the formation of hydrogen bonds (another mechanism is related to the electrostatic interactions, as discussed in Section 3.2). As far as the neutral drugs are concerned, their association with the lipid membrane discussed in Section 3.1 does not appear to be driven by hydrogen-bond formation (the average number of hydrogen bonds is not statistically significant). A more probable driving force for the latter is therefore the hydrophobic interactions of the neutral drug molecules with water, which can result in a more favorable interaction with the membrane's surface.

3.5. Diffusional Behavior

The effects of the presence of the DPPG membrane in the clustering characteristics and in the associative behavior of the drug molecules are expected to be reflected in their transport properties as well. To obtain information regarding the diffusional motion of the different molecular components of the examined system, we monitored the mean squared displacement (MSD) of their center of mass as depicted in Figure 11. For comparison purposes we have also included the diffusional behavior of the drug molecules in the membrane-free case [19].

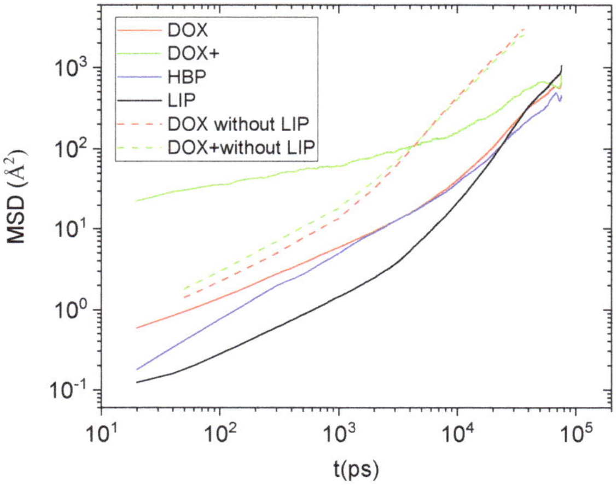

Figure 11. Mean square displacement arising from the center of mass of the drug molecules, the hyperbranched polyester, and the bilipid membrane (considered as a single moiety including all the lipids). The behavior of the drug molecules in the membrane-free case is also shown for comparison purposes.

Focusing on the behavior of doxorubicin molecules, it appears that the presence of the DPPG membrane drastically affects the drug diffusional motion. Moreover, it appears to do so to a different degree depending on whether the drug molecules are neutral or protonated. The diffusional motion of the neutral drug at all the examined timescales is slower compared to the analogous behavior in the membrane-free system, remaining very close to that of the HBP. This observation can be rationalized on one hand by the fact that the majority of the neutral drug molecules are located within the HBP's structure (see Figure 5 and Table 2) and on the other hand by the more restrictive environment (i.e., higher effective concentration compared to the membrane-free case, enhanced clustering behavior, partial association with the bilipid membrane) experienced by these molecules in the presence of the bilipid layer.

In contrast, the protonated drug molecules diffuse much faster compared to their behavior in the absence of the membrane as far as it concerns the short timescales. This trend changes at larger timescales where the transport of the center of mass of the latter becomes slower than the membrane-free case. The origin for the more complex behavior exhibited by the protonated drug should be related to the electrostatic interactions between the cationic doxorubicin molecules and the anionic part of the DPPG lipids. At short timescales, Coulombic interactions between the two oppositely charged moieties provide the driving force for an enhanced diffusive behavior. At larger timescales, where the effects of the more constrictive environment become more important, all those factors affecting the neutral drug molecules start affecting the clusters of the protonated drug molecules as well, resulting in the slowing down of their diffusional motion.

4. Conclusions

As was demonstrated by the analysis presented, the presence of the DPPG membrane imparts significant changes regarding the spatial distribution and the clustering behavior of the drug molecules. Almost 1/3 of the protonated and 1/5 of the neutral doxorubicin molecules appear to migrate from the drug/HBP complex towards the membrane's surface. Localization of drug molecules close to the membrane's surface enhances their tendency to form clusters, particularly for the positively charged drug molecules. The latter form persistent hydrogen bonds mainly with the phosphorus atoms of the hydrophilic part of the lipids. The affinity of the neutral doxorubicin molecules to the lipid bilayer appears to be driven by their hydrophobic nature. The uneven distribution of the DOX+ molecules and that of the counterions' between the HBP and the bilipid membrane results in the development of positive and negative surface charge excesses, respectively. This may eventually lead to the coalescence of the HBP/drug complex with the membrane and thus towards a mechanism for a liposomal escape in formulations in which such HBP/doxorubicin complexes are encapsulated in a DPPG-based liposomal carrier.

The timescales associated with the migration of the neutral and the protonated doxorubicin molecules from the HBP/drug complex to the membrane's surface, and thus the build-up of the final charge distribution, relates to the respective diffusional behavior. It appears that the presence of the DPPG membrane drastically affects the translational motion of doxorubicin with respect to the membrane-free case. The extent of the observed changes depends on the charge carried by the drug molecules. The neutral drug diffuses slower at all the examined timescales while the protonated drug diffuses faster at short timescales and slower at longer timescales, compared to the behavior observed in the absence of the DPPG bilayer.

The elementary mechanisms characterizing the interactions of doxorubicin molecules with a hyperbranched polymeric host in the presence of an anionic bilipid membrane may serve as a basis for a better understanding of the microscopic picture in other liposomal-based formulations for drug-delivery purposes, where similar interactions are expected to be present.

Author Contributions: Conceptualization, methodology, analysis software, writing—original draft preparation, writing—review and editing, supervision: K.K.; Data analysis, validation, visualization: P.A.; Interpretation of the results: K.K. and P.A.

Funding: This research received no external funding.

Conflicts of Interest: The authors declare no conflicts of interest.

References

1. Liu, Y.; Li, M.; Yang, Y.; Xia, Y.; Nieh, M.-P. The effects of temperature, salinity, concentration and PEGylated lipid on the spontaneous nanostructures of bicellar mixtures. *Biochim. Biophys. Acta BBA Biomembr.* **2014**, *1838*, 1871–1880. [CrossRef] [PubMed]
2. Ferber, D. GENE THERAPY: Safer and Virus-Free? *Science* **2001**, *294*, 1638–1642. [CrossRef] [PubMed]
3. Woodle, M.C.; Scaria, P. Cationic liposomes and nucleic acids. *Curr. Opin. Colloid Interface Sci.* **2001**, *6*, 78–84. [CrossRef]
4. Kontogiannopoulos, K.N.; Assimopoulou, A.N.; Hatziantoniou, S.; Karatasos, K.; Demetzos, C.; Papageorgiou, V.P. Chimeric advanced drug delivery nano systems (chi-aDDnSs) for shikonin combining dendritic and liposomal technology. *Int. J. Pharm.* **2012**, *422*, 381–389. [CrossRef] [PubMed]
5. Tsermentseli, S.; Kontogiannopoulos, K.; Papageorgiou, V.; Assimopoulou, A. Comparative Study of PEGylated and Conventional Liposomes as Carriers for Shikonin. *Fluids* **2018**, *3*, 36. [CrossRef]
6. Akbarzadeh, A.; Rezaei-Sadabady, R.; Davaran, S.; Joo, S.W.; Zarghami, N.; Hanifehpour, Y.; Samiei, M.; Kouhi, M.; Nejati-Koshki, K. Liposome: Classification, preparation, and applications. *Nanoscale Res. Lett.* **2013**, *8*, 102. [CrossRef]

7. Wenz, J.J.; Barrantes, F.J. Steroid Structural Requirements for Stabilizing or Disrupting Lipid Domains. *Biochemistry* **2003**, *42*, 14267–14276. [CrossRef]
8. Tyteca, D.; Schanck, A.; Dufrêne, Y.F.; Deleu, M.; Courtoy, P.J.; Tulkens, P.M.; Mingeot-Leclercq, M.P. The macrolide antibiotic azithromycin interacts with lipids and affects membrane organization and fluidity: Studies on langmuir-blodgett monolayers, liposomes and J774 macrophages. *J. Membr. Biol.* **2003**, *192*, 203–215. [CrossRef]
9. Li, J.; Ouyang, Y.; Kong, X.; Zhu, J.; Lu, D.; Liu, Z. A multi-scale molecular dynamics simulation of PMAL facilitated delivery of siRNA. *RSC Adv.* **2015**, *5*, 68227–68233. [CrossRef]
10. Nademi, Y.; Amjad Iranagh, S.; Yousefpour, A.; Mousavi, S.; Modarress, H. Molecular dynamics simulations and free energy profile of Paracetamol in DPPC and DMPC lipid bilayers. *J. Chem. Sci.* **2014**, *126*, 637–647. [CrossRef]
11. Kopeć, W.; Telenius, J.; Khandelia, H. Molecular dynamics simulations of the interactions of medicinal plant extracts and drugs with lipid bilayer membranes. *FEBS J.* **2013**, *280*, 2785–2805. [CrossRef]
12. Antipina, A.Y.; Gurtovenko, A.A. Toward Understanding Liposome-Based siRNA Delivery Vectors: Atomic-Scale Insight into siRNA–Lipid Interactions. *Langmuir* **2018**, *34*, 8685–8693. [CrossRef] [PubMed]
13. Tah, B.; Pal, P.; Mishra, S.; Talapatra, G.B. Interaction of insulin with anionic phospholipid (DPPG) vesicles. *Phys. Chem. Chem. Phys.* **2014**, *16*, 21657–21663. [CrossRef] [PubMed]
14. Malaspina, T.; Colherinhas, G.; de Oliveira Outi, F.; Fileti, E.E. Assessing the interaction between surfactant-like peptides and lipid membranes. *RSC Adv.* **2017**, *7*, 35973–35981. [CrossRef]
15. Ergun, S.; Demir, P.; Uzbay, T.; Severcan, F. Agomelatine strongly interacts with zwitterionic DPPC and charged DPPG membranes. *Biochim. Biophys. Acta BBA Biomembr.* **2014**, *1838*, 2798–2806. [CrossRef]
16. Habib, L.; Jraij, A.; Khreich, N.; Fessi, H.; Charcosset, C.; Greige-Gerges, H. Morphological and physicochemical characterization of liposomes loading cucurbitacin E, an anti-proliferative natural tetracyclic triterpene. *Chem. Phys. Lipids* **2014**, *177*, 64–70. [CrossRef] [PubMed]
17. Mourelatou, E.A.; Libster, D.; Nir, I.; Hatziantoniou, S.; Aserin, A.; Garti, N.; Demetzos, C. Type and Location of Interaction between Hyperbranched Polymers and Liposomes. Relevance to Design of a Potentially Advanced Drug Delivery Nanosystem (aDDnS). *J. Phys. Cem. B* **2011**, *115*, 3400–3408. [CrossRef]
18. Oberoi, H.S.; Nukolova, N.V.; Kabanov, A.V.; Bronich, T.K. Nanocarriers for delivery of platinum anticancer drugs. *Adv. Drug Deliv. Rev.* **2013**, *65*, 1667–1685. [CrossRef]
19. Karatasos, K. Self-Association and Complexation of the Anti-Cancer Drug Doxorubicin with PEGylated Hyperbranched Polyesters in an Aqueous Environment. *J. Phys. Cem. B* **2013**, *117*, 2564–2575. [CrossRef]
20. Wang, Y.; Kong, W.; Song, Y.; Duan, Y.; Wang, L.; Steinhoff, G.; Kong, D.; Yu, Y. Polyamidoamine Dendrimers with a Modified Pentaerythritol Core Having High Efficiency and Low Cytotoxicity as Gene Carriers. *Biomacromolecules* **2009**, *10*, 617–622. [CrossRef]
21. Jämbeck, J.P.M.; Lyubartsev, A.P. Another Piece of the Membrane Puzzle: Extending Slipids Further. *J. Chem. Theory Comput.* **2013**, *9*, 774–784. [CrossRef] [PubMed]
22. Jämbeck, J.P.M.; Lyubartsev, A.P. Derivation and Systematic Validation of a Refined All-Atom Force Field for Phosphatidylcholine Lipids. *J. Phys. Cem. B* **2012**, *116*, 3164–3179. [CrossRef] [PubMed]
23. Jämbeck, J.P.M.; Lyubartsev, A.P. An Extension and Further Validation of an All-Atomistic Force Field for Biological Membranes. *J. Chem. Theory Comput.* **2012**, *8*, 2938–2948. [CrossRef] [PubMed]
24. Jorgensen, W.L.; Chandrasekhar, J.; Madura, J.D.; Impey, R.W.; Klein, M. Comparison of simple potential functions for simulating liquid water. *J. Chem. Phys.* **1983**, *79*, 926–935. [CrossRef]
25. Pedretti, A.; Villa, L.; Vistoli, G. VEGA—An open platform to develop chemo-bio-informatics applications, using plug-in architecture and script programming. *J. Comput. Aided Mol. Des.* **2004**, *18*, 167–173. [CrossRef] [PubMed]
26. Wang, J.; Wolf, R.M.; Caldwell, J.W.; Kollman, P.A.; Case, D.A. Development and testing of a general amber force field. *J. Comput. Chem.* **2004**, *25*, 1157–1174. [CrossRef] [PubMed]
27. Dickson, C.J.; Rosso, L.; Betz, R.M.; Walker, R.C.; Gould, I.R. GAFFlipid: A General Amber Force Field for the accurate molecular dynamics simulation of phospholipid. *Soft Matter* **2012**, *8*, 9617–9627. [CrossRef]
28. Phillips, J.C.; Braun, R.; Wang, W.; Gumbart, J.; Tajkhorshid, E.; Villa, E.; Chipot, C.; Skeel, R.D.; Kalé, L.; Schulten, K. Scalable molecular dynamics with NAMD. *J. Comput. Chem.* **2005**, *26*, 1781–1802. [CrossRef]
29. Martyna, G.J.; Tobias, D.J.; Klein, M.L. Constant pressure molecular dynamics algorithms. *J. Chem. Phys.* **1994**, *101*, 4177–4189. [CrossRef]

30. Darden, T.; Perera, L.; Li, L.; Pedersen, L. New tricks for modelers from the crystallography toolkit: The particle mesh Ewald algorithm and its use in nucleic acid simulations. *Structure* **1999**, *7*, R55–R60. [CrossRef]
31. Po, H.N.; Senozan, N.M. The Henderson-Hasselbalch Equation: Its History and Limitations. *J. Chem. Educ.* **2001**, *78*, 1499. [CrossRef]
32. Dalmark, M.; Johansen, P. Molecular Association between Doxorubicin (Adriamycin) and DNA-Derived Bases, Nucleosides, Nucleotides, Other Aromatic Compounds, and Proteins in Aqueous Solution. *Mol. Pharmacol.* **1982**, *22*, 158–165. [PubMed]
33. Ouyang, D.; Zhang, H.; Parekh, H.S.; Smith, S.C. Structure and Dynamics of Multiple Cationic Vectors−siRNA Complexation by All-Atomic Molecular Dynamics Simulations. *J. Phys. Cem. B* **2010**, *114*, 9231–9237. [CrossRef] [PubMed]
34. Bielinska, A.U.; Chen, C.; Johnson, J.; Baker, J.R. DNA Complexing with Polyamidoamine Dendrimers: Implications for Transfection. *Bioconjug. Chem.* **1999**, *10*, 843–850. [CrossRef] [PubMed]
35. Fant, K.; Esbjorner, E.K.; Lincoln, P.; Norden, B. DNA condensation by PAMAM dendrimers: Self-assembly characteristics and effect on transcription. *Biochemistry* **2008**, *47*, 1732–1740. [CrossRef] [PubMed]
36. Lyulin, S.V.; Karatasos, K.; Darinskii, A.; Larin, S.; Lyulin, A.V. Structural effects in overcharging in complexes of hyperbranched polymers with linear polyelectrolytes. *Soft Matter* **2008**, *4*, 453–457. [CrossRef]
37. Pignatello, R.; Musumeci, T.; Basile, L.; Carbone, C.; Puglisi, G. Biomembrane models and drug-biomembrane interaction studies: Involvement in drug design and development. *J. Pharm. Bioallied Sci.* **2011**, *3*, 4–14. [CrossRef]
38. Lee, H.; Baker, J.R.; Larson, R.G. Molecular dynamics studies of the size, shape, and internal structure of 0% and 90% acetylated fifth-generation polyamidoamine dendrimers in water and methanol. *J. Phys. Chem. B* **2006**, *110*, 4014–4019. [CrossRef]
39. Chiessi, E.; Cavalieri, F.; Paradossi, G. Water and Polymer Dynamics in Chemically Cross-Linked Hydrogels of Poly(vinyl alcohol): A Molecular Dynamics Simulation Study. *J. Phys. Chem. B* **2007**, *111*, 2820–2827. [CrossRef]

MDPI
St. Alban-Anlage 66
4052 Basel
Switzerland
Tel. +41 61 683 77 34
Fax +41 61 302 89 18
www.mdpi.com

Fluids Editorial Office
E-mail: fluids@mdpi.com
www.mdpi.com/journal/fluids

www.ingramcontent.com/pod-product-compliance
Lightning Source LLC
LaVergne TN
LVHW071958080526
838202LV00064B/6779
9783039212477